The Open University

Block 5
Detail design and manufacture

Prepared on behalf of the Course Team by Steve Garner and Chris Earl with contributions by Mike Ashby and Tony Hodgson

T211 Design and designing

This publication forms part of an Open University course T211 *Design and designing*. Details of this and other Open University courses can be obtained from the Student Registration and Enquiry Service, The Open University, PO Box 197, Milton Keynes MK7 6BJ, United Kingdom: tel. +44 (0)845 300 60 90, email general-enquiries@open.ac.uk

Alternatively, you may visit the Open University website at http://www.open.ac.uk where you can learn more about the wide range of courses and packs offered at all levels by The Open University.

To purchase a selection of Open University course materials visit http://www.ouw.co.uk, or contact Open University Worldwide, Walton Hall, Milton Keynes MK7 6AA, United Kingdom for a brochure. tel. +44 (0)1908 858793; fax +44 (0)1908 858787; email ouw-customer-services@open.ac.uk

The Open University
Walton Hall, Milton Keynes
MK7 6AA

First published 2004. Third edition 2009.

Edited and designed by The Open University.

Printed and bound in the United Kingdom by The Charlesworth Group, Wakefield.

ISBN 978 0 7492 2082 2

3.1

Contents

Exercises

Introduction

This block takes us into the final stage of design – what Block 4 referred to as 'detail design'. The designer or design team might have been considering materials and processes for their product from the outset of the project but it's at this detail design stage that the specification for both must be clear and unambiguous. The materials and processes used have extensive implications for the practical success of the product (for example, its usability) and its financial viability (will it be profitable or delivered to the original budget?).

It should be clear to you by now that, in addition to the methods and procedures discussed in this course, modern design practice is being shaped by two complex and extensive forces:

- the use of computers to model, visualise and evaluate design proposals; and
- 'sustainability'.

This block provides you with a little more detail about both. Sections 1–4 deal primarily with materials, processes and manufacture. Section 5 focuses on rapid prototyping as a new means for designers to move between two- and three-dimensional models. Section 6 shows sustainability to be a combination of pressures from consumer demand, legislation, responsibilities and new opportunities. Discussion of these pressures in a block concerned with the detail design stage is not meant to suggest they can simply be addressed at the end of the design process. Quite the opposite. Both computer based working and sustainability need to be considered at the very outset of the design process. However, locating them in this block has allowed us to establish the essential framework of the design process in earlier blocks without becoming bogged down in detail.

Block 5, as with previous blocks, presents a focus on product design. Clearly this sector of the design professions is closely involved with processes of manufacture and a wide range of materials that may not form part of the domain of, say, the graphic designer. Nevertheless, the basic principles identified here could easily be identified in the practices of a wide range of design professions.

Aims and learning outcomes

Aims

1 To examine the long-standing relationship between materials, processes and mass production and to establish priorities for design activity at this stage.

2 To introduce particular examples from the world of materials and processes and to explore the profound influence these have on designed products.

3 To increase your knowledge of materials and develop skills with a particular materials database.

4 To increase your knowledge of processes and to develop knowledge and skills in 'design for manufacture'.

5 To develop understanding of the application of computers in design modelling and particularly in rapid prototyping.

6 To enable you to view various sustainability requirements within wider demands of product design and development.

Learning outcomes

As a result of studying Block 5 you should have achieved the following learning outcomes.

1 Knowledge and understanding

You should be able to demonstrate knowledge and understanding of:

1.1 the areas of knowledge required to convert a design from a layout to a manufactured item;

1.2 the broad fields of materials and their associated processes;

1.3 the key principles underpinning design for manufacture;

1.4 the exploitation of computer aided design (CAD) and the various modelling processes associated with CAD;

1.5 sustainability issues and their importance in modern design practice.

2 Cognitive (thinking) skills

You should be able to:

2.1 identify commonalities and differences between various product types in this design stage of layout to manufacture;

2.2 compare the application of particular materials and processes in a variety of common products;

2.3 identify established and more modern production systems and manufacturing technologies;

2.4 present the advantages and limitations to using various computer-based modelling tools in design;

2.4 transfer knowledge of sustainability from one application to another.

3 Key skills

You should be able to:

3.1 apply analytical skills in exploring selected examples of designed products;

3.2 apply knowledge of materials and processes from course examples to your own design activity;

3.3 demonstrate a range of IT skills including the use of computer-based data resources to assist in product analysis and design, and the construction of basic CAD models;

3.4 apply knowledge of various environmental impacts arising from design decisions to your own design activity.

4 Practical skills

You should have gained:

4.1 modelling skills in defining product detail, such as sketch modelling, computer modelling and mathematical modelling.

Study Chart

Section	Main text	Workbooks	DVD	Block 5 Supplement
1	Design for materials processes and mass production	Workbook 1	*Modelling Workbook* Activity 20	Study guide
2	Materials and product design			
3	Materials and processes		CES Selector (software)	CES materials and processes database (Lessons 1–3)
4	Processes: the manufacturing spectrum		CES Selector (software)	CES materials and processes database (Lessons 4–5)
5	Computer modelling for design		solidThinking (software)	solidThinking: making simple CAD models (Lesson 3)
6	Sustainable design		Eco-it (software) Philips: Design development and sustainability (video sequence)	ECO-it: evaluating sustainability

Support notes for TMA 05 appear in *Block 5 Supplement*. You should refer to these as you prepare to answer TMA 05.

1 Design for materials, processes and mass production

Figure 1 Michael Thonet Chair No. 14, 1859

To begin this section I have chosen to return to an artefact you first came across in Block 1 – the Thonet (pronounced *thon-ay*) chair, shown here in Figure 1. It provides a good opportunity to make a few introductory observations about the relationship between choice of materials, detail design, materials processing, manufacture and assembly before we move into the somewhat greater complexities of the mass production of products today.

1.1 Gebrüder Thonet

1.1.1 The Thonet chair and its component parts

In Block 1 (Section 5.2), the Thonet chair provided a useful illustration of the technique of steaming and bending beech wood to make furniture. At first sight, the Thonet chair seems to exemplify traditional hand-produced furniture.

Yet the Austrian, Michael Thonet (1796–1871), and his brothers developed the first chairs to be produced in very large quantities to standardised designs. Their designs were based upon techniques of steam bending wood, invented and patented by Michael Thonet in the mid 19th century. The cross-section of the wood was round and of small diameter (typically 20–30 mm), and steaming gave a range of long curved shapes. The bentwood components were then screwed, bolted and glued together using 'batch production' techniques.

Thonet chairs were utilitarian, unpretentious and cheap. They became the standard furniture for Viennese cafés and later for French and German interiors. They were perceived as outside the culture of 'artistic' taste, yet became commonplace throughout Europe and remain so today. Some of Thonet's designs, with only small variations, continue to be produced.

These designs were first popular as non-domestic furniture. In the mid-1920s, the Swiss architect Charles Edouard Jeanneret (known by his *nom de plume* 'Le Corbusier') designed and built a handful of revolutionary houses. The furniture he selected for these houses was that of the Thonet company. One of his favourite designs, model B-9 (Figure 2), has been renamed after him and is sometimes mistakenly said to be his design. Whether attributed to this designer or not, his use of the chair means that many people associate it with the 1920s. Yet this particular chair was designed in 1870. Moreover, the production techniques which led to this chair were introduced 20 years earlier. All this occurred several decades before Le Corbusier was born. He was attracted to the Thonet chairs by their austerity, utility and simplicity. From their outward appearance, we might think of them as exceptionally simple. But is their construction really so simple? Let us consider the parts and assemblies of a typical Thonet chair (as we did for the IKEA chair in Block 1).

Figure 2 Thonet B-9 chairs in Le Corbusier interior at Esprit Nouveau Pavillion 1925

Look at the parts of the Thonet chair, design No. 14 of 1859. Figure 3 shows six bentwood components, eight screws and two nuts and bolts. The seat is usually made of woven cane or a piece of plywood, steam-embossed with a pattern.

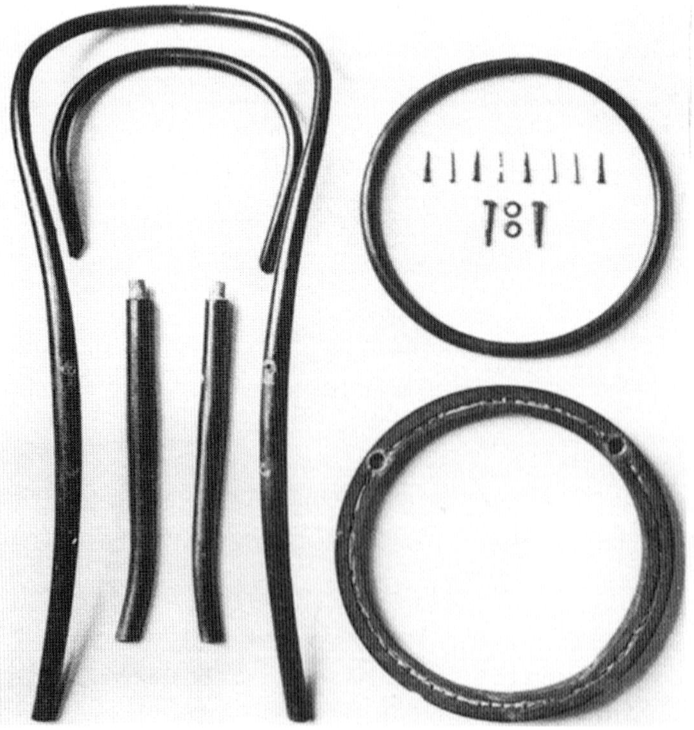

Figure 3 Components of the Thonet chair

I have a similar chair at home, which has additional side braces between the back and seat. I examined this chair carefully. My notes are reproduced opposite and my sketch appears in Figure 4 (overleaf).

Example Notes on the Thonet Chair

Components

1 Back and rear legs in one piece
2 Front leg, left
3 Front leg, right
4 Seat frame
5 Back brace
6 Side brace, left
7 Side brace, right
8 Underseat circular brace
9 Plywood seat

Fixings and fastenings

There are also:

- 2 wooden brackets, which clamp each front leg to the seat
- 2 bolts with nuts and washers
- 4 large screws
- 12 medium screws
- 2 dowels

The chair weighs about 3.5 kg.

Materials and processing

The wood is beech, steamed and bent around formers.

Some machining has rounded the ends of the braces and back support.

The seat frame has had a 5 mm x 15 mm wide rebate cut into it. This rebate accepts the plywood seat.

This frame, although 35 mm in section, has been steam formed into a circle of 360 mm diameter. The circle is closed by tapering, tailoring and dowelling the ends.

Similarly, the circular leg brace is tapered and joined.

Assembly

The U-shaped piece that forms the back and rear legs is bolted to the seat frame by two 70 mm long bolts, tightened with washers and nuts.

The front legs are glued into their slots in the seat frame, located by dowels and fixed, by two long curved wooden brackets, using both screws and glue.

Each side brace is screwed to the seat and back at four points.

The back brace is similarly fixed at four points to the back.

The seat of plywood is inserted in the frame and fixed only by glue.

I believe the chair has been assembled approximately in the order indicated here, the first operations taking place with the chair inverted and the last three the correct way up.

Unknowns

Such is the quality of the workmanship that it is hard to see the join in the circular seat brace. But I know wood does not grow into a circle. Small dowels pin the seat frame into its circular form. No such dowels are visible on the underseat brace.

The fixings between the back and the back brace are disguised by countersinking and stop dowels. The dowels may go all the way through but I would have expected them to work loose by now if that was the case. These dowels are carefully shaped and polished to the contour of the back brace. I think this would have been the last assembly operation.

Comments

There is a difference in quality between the visible fixings and the underside fixings. The screws to the side braces are round-headed and countersunk. The larger screws fixing the underseat brace to the legs are neither, and actually protrude.

The chair is an original Thonet rather than a recent Czech import. It has been painted then stripped by immersion in solvent at some time. This process loosens the old glues. Occasionally the chair develops a small lateral twist and a squeak, but this can be remedied by tightening the underseat screws.

Exercise 1 Identifying components and sketching them

Figure 4 shows my sketch of the components of the Thonet chair. Select any simple chair that you can find and examine it. Try to identify the separate components and produce a similar style of perspective sketch to the one in Figure 4 using your selected chair. Notice that in Figure 4 the components appear to be 'exploded' apart, including the screws and other fixings.

> Perspective sketching that conveys component parts as if a product has been exploded apart is covered in Section 7 of the *Modelling Workbook*.

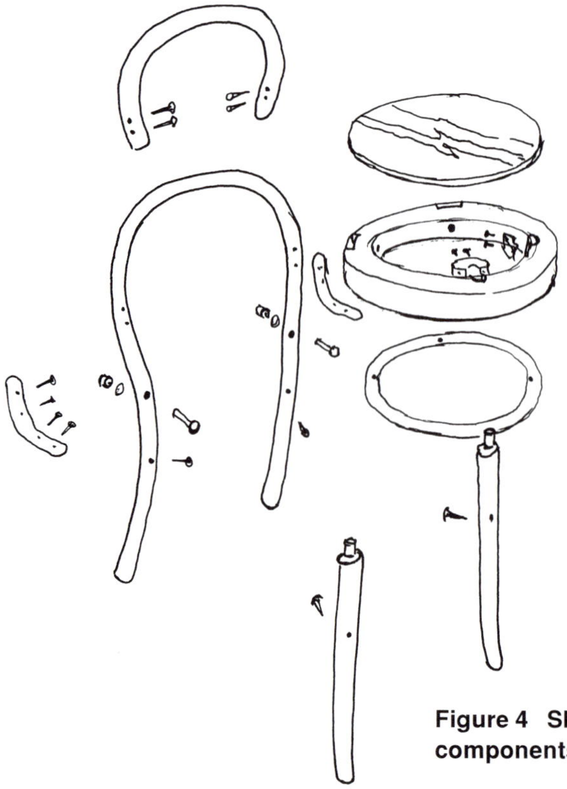

Figure 4 Sketch of the Thonet chair components in exploded form

1.1.2 The Thonet manufacturing process

From such an inspection of a Thonet chair it is not possible to deduce all of the processes of manufacture, so here is a brief outline.

Materials selection

The usual material for a Thonet chair was the common beech, a straight-grained tree growing plentifully in the forests of Moravia. The wood is finely and evenly textured and, with a moisture content of 35%, lends itself to staining and polishing. Once felled, the trees had to be used within six months to avoid the loss of natural resins essential to the bending process. Factories were located near the forests to minimise haulage.

Sawmills processed the logs into different lengths and widths corresponding to specific furniture components. A proportion of wood was turned on a lathe to give round sections. This particular design of lathe was specially devised and developed by Thonet himself. Approximately 40% of the trunk was brought to the factory in a processed form.

The process of making Thonet chairs is illustrated in Figure 5.

Sawing logs into sections

Lathe turning round sections

Steaming

Bending the softened wood to formers

Finishing components on a sanding belt

Protecting and packing components

Figure 5 Historic photographs of the Thonet factory and its processes

Steaming

The turned wooden rods and other sections were air-dried, then steamed in a retort (a pressure vessel). The length of time each batch of components was steamed depended on the moisture level and cross-section of the individual parts, and the batch size, but it could be up to several hours. Once steamed, each length was tested for weakness and imperfections.

Bending

Bending had to occur immediately after steaming and checking, while the sections were still hot and pliable. Steel straps were attached and clamped to the wooden rods. Scrap veneer was sandwiched between the strap and the rod to protect the wood. The metal straps followed the convex curves of the design, then the whole sub-assembly was bent into a rigid cast-iron former (mould) which took up the concave curvatures (Figure 6). This procedure required two or three strong workers.

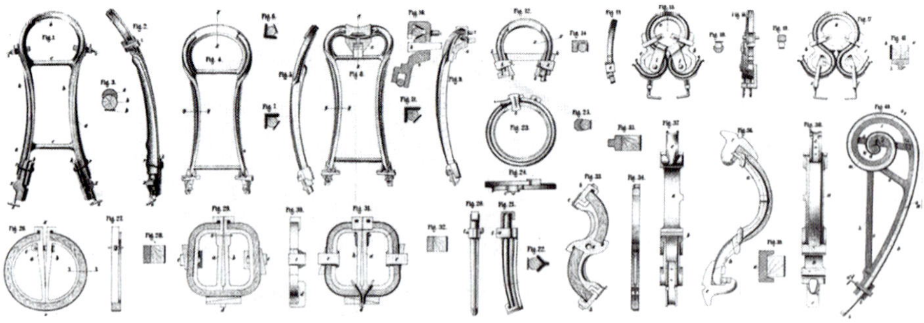

(a) Illustration of some of Thonet's formers

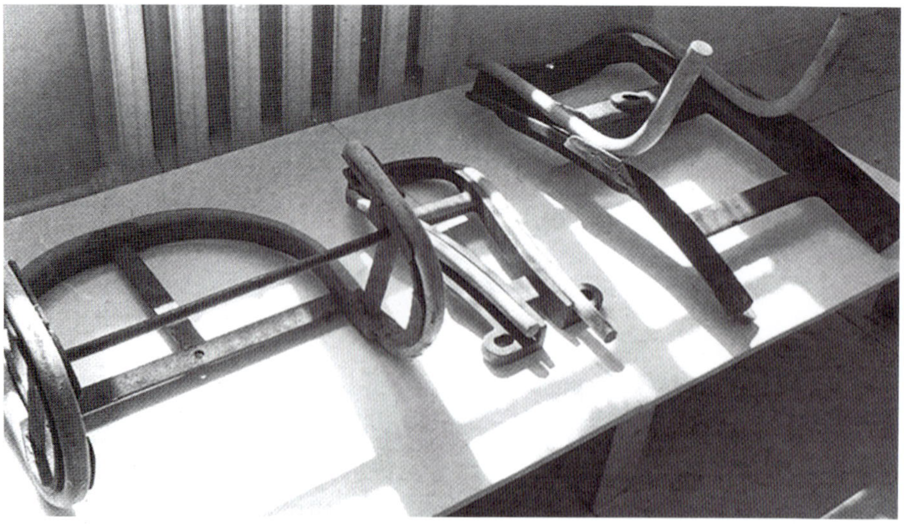

(b) Iron formers and steel straps used in bending the beech rods in Thonet chairs

Figure 6

Drying

The filled formers (or moulds) were placed in warm drying rooms at 40 °C where the moisture level stabilised overnight.

Machining

The pieces were then removed from the formers, which were replaced by temporary wooden stretchers. The pieces were machined to accept screws and bolts to fit to other members. Once removed from the wooden stretchers, they were sanded with various grades of abrasive paper and stained.

Assembly

Some pieces, such as the backs and the seats, were fitted with additional pieces or were caned. Chairs were then screwed and glued together, then wrapped in paper for dispatch.

However, some articles of furniture were delivered in pieces for local assembly by the retailer. In the words of the first catalogue of 1859, 'A case with 3 dozen chairs when they are taken to pieces occupies a space of 36 cubic feet.' (Figure 7).

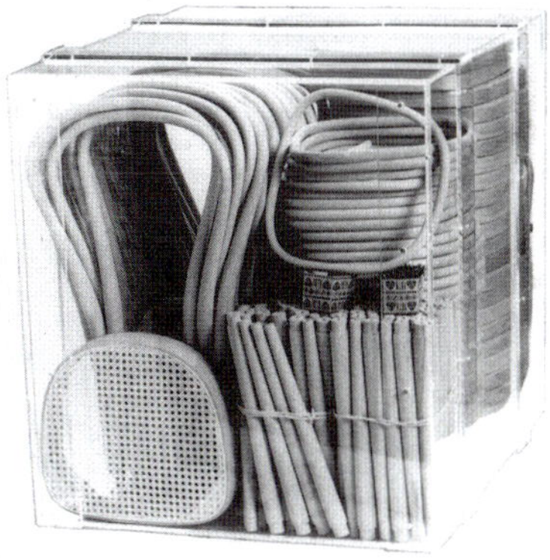

Figure 7 Transparent one-metre cube showing the packing of two dozen demounted chairs

1.1.3 Discussion: simplicity and complexity

So is the Thonet a simple chair? At one level perhaps it is – in its structurally efficient use of material and its general form. But examining the chair more carefully, for example through the process of drawing it, reveals that the main pieces form quite complex curves in two planes and the round cross-sections alter along the length of these curves. If we were to generate these forms using our CAD program then the mathematical description of the curves would be far from simple. More than that, closer scrutiny reveals that the preparation of the raw material, the steam processing, the formers for bending the wood, the fixings and the assembly of the components are not simple either.

Michael Thonet was an inventor, but he invented a *manufacturing process* rather than a specific design. His experiments with laminated wood had led him towards the idea that whole sections of wood could be steam bent. He devised the machines and equipment that made the process work. He invented a specific form of lathe for dealing with long sections. With his associates, he devised the complex cast-iron formers and straps, which moulded the wood in its plastic (permanently deformable) state. These formers and straps overcame the key problems of bending wood: its tendency to split on its outer surfaces and buckle on its inner surfaces.

Furthermore, the organisation of material and the process made a large range of designs possible (Figure 8). The Thonet catalogues show the enlargement of the range from 25 models in 1859 to over 1200 models by 1911 (Candilis *et al.*, 1980, p.15). These numbers give a misleading impression of the diversity of Thonet's output. Like other manufacturers, the range was built up by creating related variants from a relatively small number of standard parts.

This was not mass production in the modern sense. The Thonet process was still tied to *batch production*, where the work moved from workshop to workshop in batches, rather than the *flow-line production* which became the dominant characteristic of the motor car industry from the time of Henry Ford onwards. These production methods are discussed later in the block.

However, Thonet's process still advanced thinking about mass manufacture. He simplified production with pre-processing of materials, standardisation, use of interchangeable parts, division of labour, and the packaging of furniture in parts for local assembly. All these elements are now thought of as essential to modern mass production using flow-lines.

Batch production did not prevent the company from generating a massive output. By 1911 the company had produced over 50 million No. 14 chairs (according to its own conservative estimate in its catalogue of that date). By 1913 it was producing 1.8 million items of furniture per year, of which 1.3 million were chairs (Wilk, 1980).

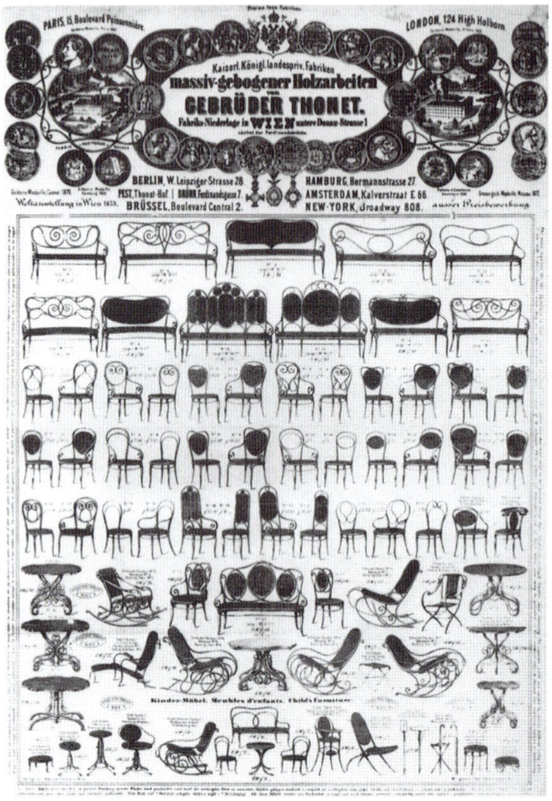

Figure 8 Some of the range of Thonet furniture products available towards the end of the 19th century

SAQ 1

What are the main characteristics of Thonet chairs which we can think of as belonging to modern mass production?

1.2 Specifying dimensions for manufacture

The success of the Thonet chair can be partly attributed to clever use of standardisation and interchangeability. To obtain these characteristics one of the things a designer needs to do is to pay special attention to specifying the dimensions of components. This is as true today as it was for Michael Thonet.

Interchangeability of parts was a key step in industrialisation, which was first seen in the manufacture of, for example, clocks and guns in the 18th century. However, in gun manufacture and that of products such as sewing machines, typewriters and early cars, interchangeability could only be achieved by a greater precision than was required in the manufacture of products such as furniture.

Figure 9 Lego Duplo blocks

At the detail design stage the designer must think ahead to how the component will be manufactured. Assigning dimensions to features as we did in Block 4 is not sufficient by itself to ensure standardisation and interchangeability.

Suppose, for example, we want to make a block for a toy construction kit such as Lego. Figure 9 shows coloured blocks from the Lego Duplo range for young children. Later in this block you will find out a little more about these blocks – what material they are made from (ABS plastic) and how they are manufactured (injection moulding) – but for the moment I concentrate on how a designer specifies their dimensions.

> Abbreviations and acronyms for materials, such as ABS, are used throughout this block for ease of study. Their full names are given in Appendix A.

Let me talk you through the dimensions of the 'double' Duplo block. The measured dimensions are: length 63 mm, width 31.5 mm and height 19 mm. The top of the block has eight, regularly spaced hollow circular projections. These have outer radius 4.5 mm and inner radius 3.5 mm. The distance between centres of the circular projections is 16 mm and their height is 4 mm. On the underside of the block is a complicated hollow cavity with three circles and several short projections from the outer wall (see Figure 10). The wall thicknesses of the block are 1 mm.

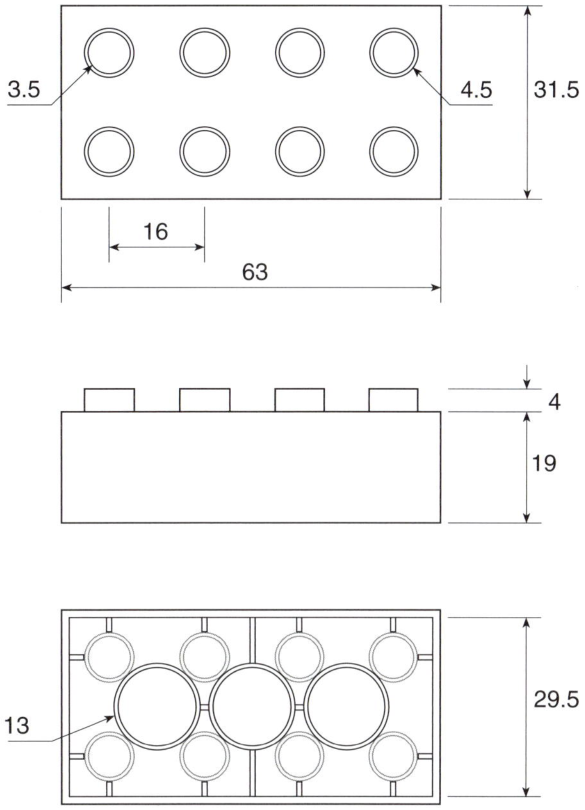

Figure 10 Drawing of a Lego Duplo block

Dimensions in millimeters.

The eight circular projections on the top of one block fit into the spaces in the base of another. The positions are shown in the view of the underside of the block. Note that there are three points at which each circular projection touches as it fits into the base of another block.

There are two things I want you to consider:

1 No manufacturing process produces parts to exactly the right dimensions. All processes yield components with some slight variations in their dimensions. Sometimes these variations are very small; nevertheless they do exist and they are important.

2 The toy blocks must always fit together snugly. This is a user requirement. It should not require too much force to assemble two blocks together and the blocks should click together holding firmly. If they are too loose then constructions would fall apart; if too tight they would be difficult to assemble and disassemble. This requirement becomes even more onerous when we think of a complex construction with many interlocking blocks. Wherever we add a new block it must fit snugly to its neighbours.

What would happen if the blocks were manufactured so that there was a wide variation in dimensions? Then the second requirement might not be met. Projections on the top of one block might be too big for the holes in the base of another block (preventing them fitting together or making it difficult to disassemble them). If the projections are too small then the loose fit means that the blocks might not hold together.

In order to meet the user requirements that all blocks fit snugly the designers of the blocks need to specify the variations allowed in the dimensions during manufacture. Exactly how they determine these allowable variations is complicated and outside the scope of this course but it largely concerns the materials and the processes chosen. The specified range of variation should be achievable for the process used to make the component. If the range is too small it may be impossible or too expensive to manufacture.

For a particular process (such as the injection moulding of the plastic used in the Lego Duplo blocks) there will be a typical or recommended variation for the process as well as a lower limit of variation which is achievable for that process.

tolerance
The allowable variation in a specified size.

In preparing a design for manufacture each dimension is specified by a smallest dimension and a largest dimension. The difference between these two dimensions is called the tolerance. When designers specify tolerances they are estimating how precisely a component needs to be made to deliver its intended function. However, they should balance this with difficulties of manufacturing the component to that tolerance. A narrow tolerance may be prohibitively expensive to manufacture leading the designer to consider changing the design.

It might be necessary to specify a very small variation in say, a plastic component of a drug infusion pump where performance is vital, but we should question whether this level of accuracy and cost should be built into a child's toy. For the Lego block made out injection moulded plastic the recommended variation on the dimensions achievable by manufacture is about 0.1 mm. Reducing the variation beyond this recommended level leads to steep increases in manufacturing costs. Box 1 looks in more detail at the application of tolerances in the Lego block example.

Box 1 Applying tolerances to the Lego Duplo block

A straightforward way to specify the allowable variation for a particular dimension is to give the range of dimensions; that is the largest and the smallest allowable dimensions. So the length of the Lego block, with a recommended variation of 0.1 mm, might be given as lying in the range between 62.95 mm and 63.05 mm. The single dimension on the component drawing (63 mm in this case in Figure 10) is a guide or 'nominal' dimension. Indeed the range can be specified by the distances between this 'nominal' dimension and each of the limits. So the range for the 'nominal' 63 mm length with a variation of 0.1 mm is specified as lying in the range between 63 - 0.05 mm and 63 + 0.05 mm or 63 ± 0.05 mm. We have assumed here that the range is 0.1 mm, the typical variation for the material (ABS) and process (injection moulding).

The actual dimensions of manufactured blocks should all lie in the specified range. Let's look at the dimensions for the circular projections on the top of the block and the holes in the underside. To ensure a snug fit the projections can be very slightly larger than the holes since the plastic material of the blocks gives a little. The projection might range in diameter between 9.0 mm and 9.1 mm whilst the holes range in 'diameter' from 8.9 mm to 9.0 mm. We have again assumed that the range is 0.1 mm, the typical variation for the material (ABS plastic) and process (injection moulding). Because there are eight projections on each block the spacing of the projections and corresponding holes is also critical for the blocks to fit together. Specifying maximum and minimum dimensions for all the features of Lego blocks is a complex task because of the requirements that blocks fit together.

A key challenge for manufacturing engineers is to develop manufacturing methods where tolerances can be controlled. If parts are going to be assembled together like the Lego blocks they must all be made within their designed tolerances otherwise the assembly will not function properly. For mass-manufactured parts today in plastics and metal – just as for Thonet's chair parts in wood – this means that the manufacturing variations on possibly millions of repeats of the same part must all be the same. The range of variation certainly must not increase during a long manufacturing run. Why might the range of variation increase? Take another look at the Lego block. It is made in an injection moulding machine where plastic granules are melted and forced into a steel mould to give the shape. The mould is made in several pieces, which are separated to release the final component. Over time these moulds will gradually wear, so that the manufactured blocks will change dimensions very slightly. This will increase overall variation. There comes a point at which the moulds become so worn that the variations start to go outside the designed tolerance. The moulds then need to be replaced.

These moulds are expensive to make, so a manufacturer will not want to replace them unnecessarily. Hard materials are, therefore, used for the moulds and sometimes their surfaces are hardened or coated to reduce wear. The moulds themselves must be manufactured to even narrower tolerances than the injection moulded parts being made. High tolerance moulds are expensive to make especially in hard materials, which resist wear.

In the case of the Lego block we have an example where parts are made accurately with small tolerances. This is necessary so that any mixture of blocks can be fitted together snugly. The mould tools are thus expensive to make and particular attention is given to making the surfaces very hard, with special coatings, so that they wear slowly.

Designing appropriate tolerances

A major task of manufacturing is to control tolerances in the manufactured parts so that they match the tolerances specified by the designer. The designer shoulders a heavy responsibility here. If they specify tolerances that are unnecessarily narrow, then manufacturing will be more expensive than it need be, jeopardising the commercial success of a product.

This part of the course is about the inter-relationship of materials and manufacturing processes. It is about the decision-making process that has to wrestle with these broad responsibilities, trying to exploit the best of each, within constraints that might be practical or financial. Thus a broad knowledge base is required by designers – including forming, shaping, manufacturing and even assembling. The brief examinations of Thonet and the Lego block reveal the decision-making process to be far from clear. Factors are traded off against one another. Perhaps a component can be made slightly cheaper if a different material is specified or if one accepts a slightly lower tolerance? Perhaps clever design can reduce the need for narrow tolerances thus opening the way to use cheaper manufacturing techniques or cheaper materials? At this stage the design team will rely on the expertise of the material scientists and the manufacturing engineers to provide the necessary information. However, this shouldn't be viewed as a process of handing over the problems for someone else to deal with. The decisions arising at the detail design stage are the direct result of decisions made at the specification, concept and layout stages, and the design team need to follow these through to manufacture. Also, there is plenty of scope at this stage for creative thinking, particularly where new problems in manufacturing or assembly need to be resolved quickly or new technical opportunities present themselves. For many designers this is a very rewarding stage in the design process. Not only are designs converted from prototypes to manufactured items but by being involved in the process the designer can add to their own bank of knowledge, which in turn will improve their ability to create more practical or inventive solutions to future concept design briefs.

Key points of Section 1

Examining and evaluating a design involves the designer in the detail of parts, materials and processes as well as the overall form.

The properties of materials and processes are some of the major influences on the shape of parts and therefore on a product as a whole.

The Gebrüder Thonet example illustrates the importance of designing an efficient production process as well as designing an effective product. Michael Thonet was as creative with manufacturing as he was with the style and form of his chairs. He also understood the properties of the materials he worked with.

Thonet exploited a number of important techniques that were to become central to efficient mass manufacture in the 20th century. The Thonet company wasn't the first to exploit these but they did successfully integrate them and became hugely successful in the field of furniture production. The techniques included: pre-processing of materials, interchangeable parts, division of labour and distributing products as parts for assembly elsewhere.

The resulting products, such as Thonet's chairs, can appear deceptively simple when in fact they may be far from simple, involving the integration of a complex range of components.

One of the key activities for a designer is to accurately specify the dimensions of components. In doing this the designer will have to determine the acceptable tolerances in dimensions. These are influenced by the properties of the materials and processes that the designer wishes to use as well as requirements to do with usability or aesthetics.

(Learning outcomes 1.1, 2.1, 3.1, 4.1)

W 1

Now would be a good time to complete Activity 20 in the *Modelling Workbook* if you haven't already done so.

2 Materials and product design

Mike Ashby

2.1 Introduction

Good design works. Excellent design also gives pleasure.

Products achieve success through a combination of sound technical design and imaginative industrial design. The amalgam creates product character – the way materials and processes are used to provide functionality, usability and satisfaction in ownership. This last – satisfaction – is greatly influenced by the aesthetics, associations and perceptions that the product carries. This combination can be called the product personality.

product character
The distinctive nature of a product imparted by, for example, its form or the materials and processes used in its creation.

product personality
The combination of aesthetics, associations and perceptions that a product evokes for its user.

Materials, you might say, are the foodstuffs of design; processes the ways of preparing the materials for consumption. How are they chosen? The materials and processes must meet the technical requirements of the design, establishing its functionality, its safety and its cost. These are mainly the concern of the technical engineer. Here we focus on another set of criteria, equally important in product design: how do designers choose a product's materials and processes to create its *personality*? This question does not require a degree in engineering to answer – personality, after all, is a human quality. What it does require is an ability to observe, to compare and to relate.

This essay is about the way materials and processes can be used to create product personality and character.

2.2 The requirements pyramid

The pen with which I am writing cost £5 (Figure 11, upper images). If you go to the right shop, you can find a pen that costs well over £1000 (lower images). Does it write 200 times better than mine? Unlikely; mine writes perfectly well. Yet there is a market for such pens. Why?

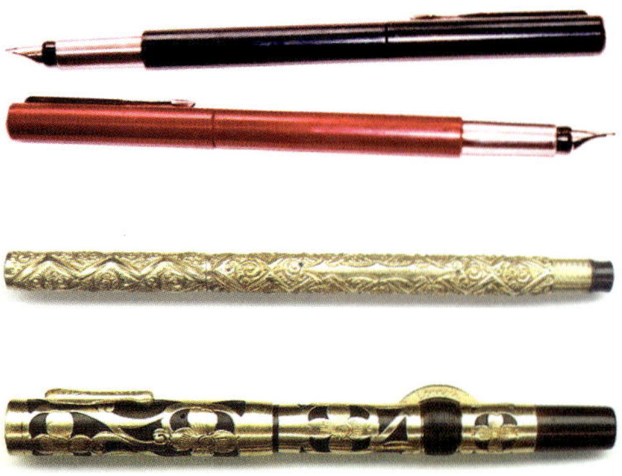

Figure 11 Pens, inexpensive and expensive

The chosen materials – acrylic in the upper two images and rolled gold in the lower ones – create the aesthetics and the associations of the pens. They are perceived differently, one pair as utilitarian, the other as something rare and crafted. The perceptions have much to do with the materials of which they are made.

As you saw in Block 2, product has a *cost* – the outlay in manufacturing and marketing it. It has a *price* – the sum at which it is offered to the consumer. And it has a *value* – a measure of what the consumer thinks it is worth. The expensive pens command the price they do because the consumer perceives their value to justify it. What determines value?

Functionality, provided by sound technical design, clearly plays a role. The requirements pyramid of Figure 12 has this as its base: the product must work properly, be safe and economical. Functionality alone is not enough: the product must be easy to understand and operate, and these are questions of *usability*, forming the second tier. The third, completing the pyramid, is the requirement that the product gives *satisfaction*: that it enhances the life of its owner.

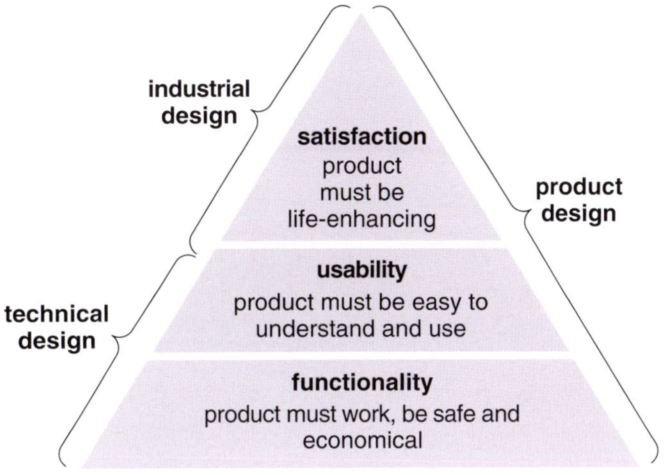

Figure 12 The requirements pyramid

The lower part of the pyramid tends to be called 'technical design', the upper part 'industrial design', suggesting that these are separate activities. It is better to think of all three tiers as part of a single process called 'product design'.

The value of a product is a measure of the degree to which it meets (or exceeds) the expectation of the consumer in functionality, usability and satisfaction. Think of this as the character of the product. It is very like human character. An admirable character is one who functions well, interacts effectively and is rewarding to be with. An unappealing character is one that does none of these. An odious character is one that does one or more of these in a way so unattractive that you cannot bear to be near them.

Products can share characteristics. All the pens in Figure 11 function well and are easy to use. The huge difference in price implies that the lower pair provides a degree of satisfaction not offered by the upper ones. The most obvious difference between them is in the materials they consist of – the upper pair of moulded acrylic, the lower pair of gold. Acrylic is the material of toothbrush handles, something you throw away after use. Gold and silver are the materials of precious jewellery – they have associations of craftsmanship, of heirlooms passed from one generation to the next. But how do designers create character and value without using precious materials?

2.3 Product character

Figure 13 overleaf shows one way of dissecting product character. It is a map of the ideas we are going to explore; like all maps there is a lot of detail, but we need it to find our way. In the centre is information about the *product* itself: the basic design requirements, i.e. its function and its features. The way this is approached and developed is conditioned by the *context*. The context is set by the answers to the questions: *Who? Where? When? Why?* Consider the first of these: *Who?* A designer seeking to create a product attractive to women will make choices that differ from those for a product intended for children, or for elderly people, or for sportsmen. *Where?* A product for use in the home requires a different choice of material and form than one to be used in a school or hospital. *When?* A product intended for occasional use is designed in a different way to one that is used all the time; one for formal occasions differs from one for informal occasions. *Why?* A product that is primarily utilitarian involves different design decisions than one that is largely a life-style statement. The context influences and conditions all the decisions that the designer takes in finding a solution. It sets the *mood*. You saw the use of mood boards in Block 2. Many designers, working on a project, assemble a mood board with images of the sort of people for whom the product is intended, the surroundings in which they suppose it will be used, and other products that the intended user group might own, seeking to capture the flavour of their lifestyle.

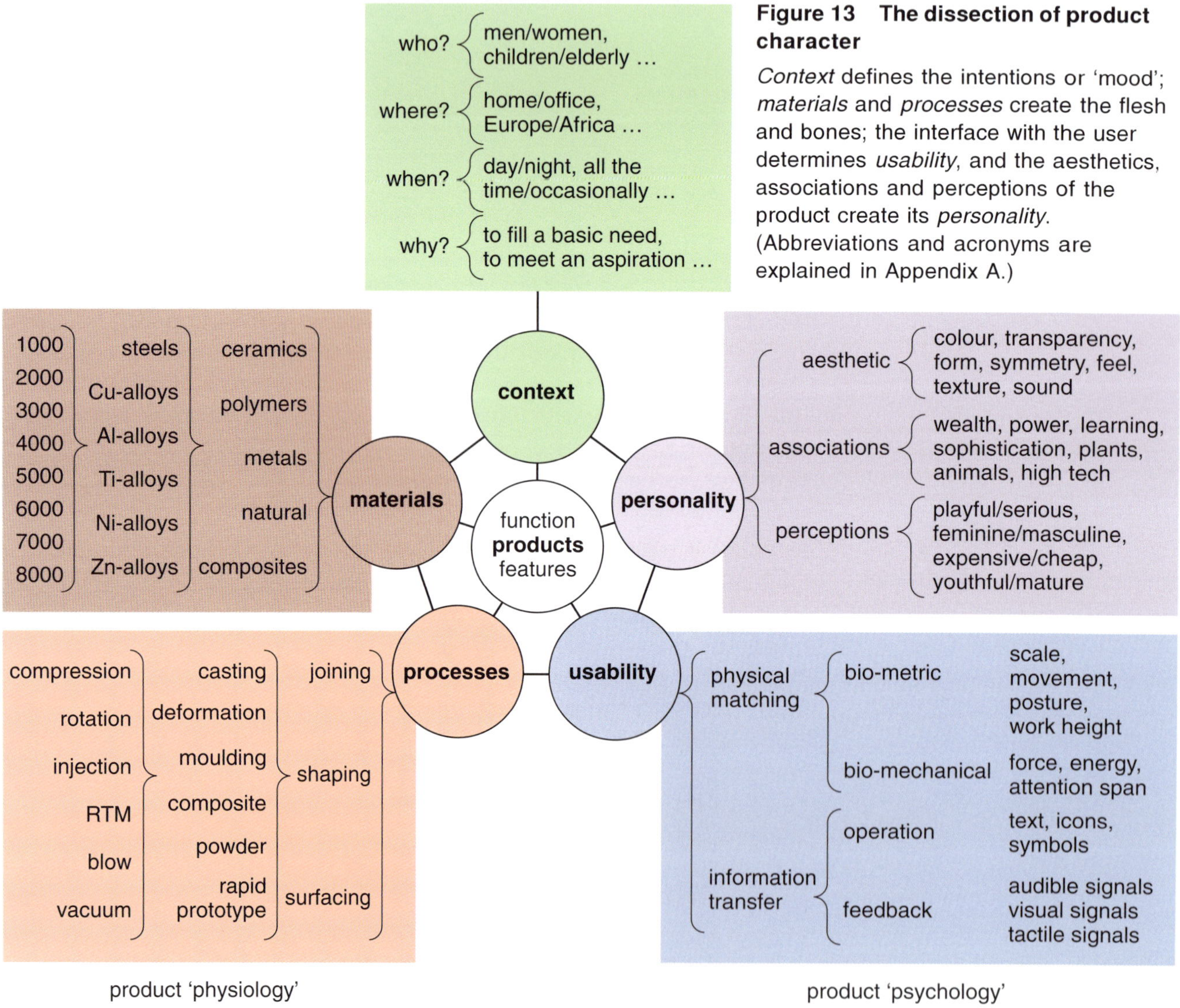

Figure 13 The dissection of product character

Context defines the intentions or 'mood'; *materials* and *processes* create the flesh and bones; the interface with the user determines *usability*, and the aesthetics, associations and perceptions of the product create its *personality*. (Abbreviations and acronyms are explained in Appendix A.)

On the left of products in Figure 13 lies information about the *materials* and the *processes* used to shape, join and finish the product – the main focus of this essay. Each illustrates the menu, so to speak, from which choices can be made. The primary step in selecting both material and process is that they can meet the constraints imposed by the primary design requirements – the essential functions and features of the central circle. Material and process give the product its tangible form, its flesh and bones so to speak; they create the *product physiology*.

On the right are two further packages of information. The lower one – *usability* – characterises the ways in which the product communicates with the user: the interaction with their sensory, cognitive and motor functions. A product's success requires a mode of operation that, as far as possible, is intuitive, does not require much effort, and an interface that communicates the state of the product and its response to user action by visible, acoustic or tactile responses. It is remarkable how many products fail in this, and by doing so, exclude many of their potential users.

One circle remains: the one labelled *personality*. Product personality derives from *aesthetics*, *associations* and *perceptions*.

Anaesthetics dull the senses. Aesthetics do the opposite: they stimulate the five senses of sight, hearing, touch, taste, smell. We are concerned here with colour, form, texture, feel, smell and sound – think of the smell of a new car and the sound of its door closing.

Products also have associations: the things they remind you of, the things they suggest. The Land Rover and other sports utility vehicles (SUVs) have forms and (often) colours that mimic those of military vehicles. The streamlining of American cars of the 1950s and 1960s carried associations of aerospace. It may be an accident that the VW Beetle has a form that suggests the insect, but the others are no accident; they were deliberately chosen by the designer to appeal to the consumer group (the *Who?*) at which the product was aimed.

Finally, the most abstract quality of all, perceptions. Perceptions are the reactions the product induces in an observer, the way it makes you *feel*. Here there is room for disagreement; the perceptions of a product change with time and depend on the culture and background of the observer. Yet in the final analysis it is the perception that causes the consumer, when choosing between a multitude of similar models, to prefer one above the others; it creates the 'must have' feeling. Table 1 lists some perceptions and their opposites, in order to sharpen the meaning. They derive from product reviews and magazines specialising in product design; they are a part of a vocabulary, one that is used to communicate views about product character (aesthetics, associations and perceptions are discussed more fully in *Materials and Design* by Ashby and Johnson, 2002).

Table 1 Some perceived attributes of products and their opposites

attribute	opposite	attribute	opposite
aggressive	passive	extravagant	restrained
cheap	expensive	feminine	masculine
classic	trendy	formal	informal
clever	silly	handmade	mass produced
clinical	friendly	honest	deceptive
common	exclusive	humorous	serious
decorated	plain	informal	formal
delicate	rugged	irritating	loveable
disposable	lasting	lasting	disposable
dull	exciting	mature	youthful
elegant	clumsy	nostalgic	futuristic

2.4 Using materials and processes to create product personality

Do materials have a personality? There is a school of thinking that holds as a central tenet that materials must be used 'honestly'. By this they mean that deception and disguise are unacceptable – each material must be used in ways that expose its intrinsic qualities and natural appearance. It has its roots in the tradition of craftsmanship – the potters' use of clays and glazes, the carpenters' use of woods, the skills of silversmiths and glass makers in crafting beautiful objects that exploit the unique qualities of the materials with which they work – an integrity to craft and material.

This is a view to be respected. But it is not the only one. Design integrity is a quality that consumers value, but they also value other qualities: humour, sympathy, surprise, provocation, even shock. You don't have to look far to find a product that has any one of these, and often it is achieved by using materials in deceptive ways. Polymers (commonly called 'plastics') are frequently used in this way – their adaptability invites it. And, of course, it is partly a question of definition – if you say that a characterising attribute of polymers is their ability to mimic other materials, then using them in this way *is* honest.

2.4.1 Materials and the senses: aesthetic attributes

Aesthetic attributes are those that relate to the senses: touch, sight, hearing, taste and smell (Table 2). Almost everyone would agree that metals feel 'cold'; that cork feels 'warm'; that a wine glass, when struck, 'rings'; that a pewter mug sounds 'dull', even 'dead'. A clear polystyrene water cup can look indistinguishable from one made of glass, but pick it up and it feels lighter, warmer, less rigid; tap it and it does not sound the same. The impression it leaves is so different from glass that, in an expensive restaurant, it would be completely unacceptable. Materials, then, have certain aesthetic attributes that characterise them. Let us see if we can pin these down.

Table 2 Aesthetic attributes associated with each sense

Sense	Attribute
touch	warm cold soft hard flexible stiff
sight	optically clear transparent translucent opaque reflective glossy matt textured
hearing	muffled dull sharp resonant ringing low pitched high pitched
taste/smell	bitter sweet

Touch: soft–hard/warm–cold

Steel is 'hard'; so is glass; diamond is harder than either of them. Hard materials do not scratch easily; indeed they can be used to scratch other materials. They generally accept a high polish, resist wear and are durable. The impression that a material is hard is directly related to the material property *hardness*, measured by materials engineers and tabulated in handbooks. Here is an example of a sensory attribute that relates directly to a technical one.

'Soft' sounds like the opposite of 'hard' but, in engineering terms, it is not – there is no engineering property called 'softness'. A soft material deflects when handled, it gives a little, it is squashy, but when it is released it returns to its original shape. Elastomers (rubbers) feel soft; so do polymer foams, and this has to do with the engineering property *elastic modulus*: both have moduli that are 100 to 10,000 lower than ordinary 'hard' solids; it is this that makes them feel soft. Soft to hard is used as one axis of Figure 14.

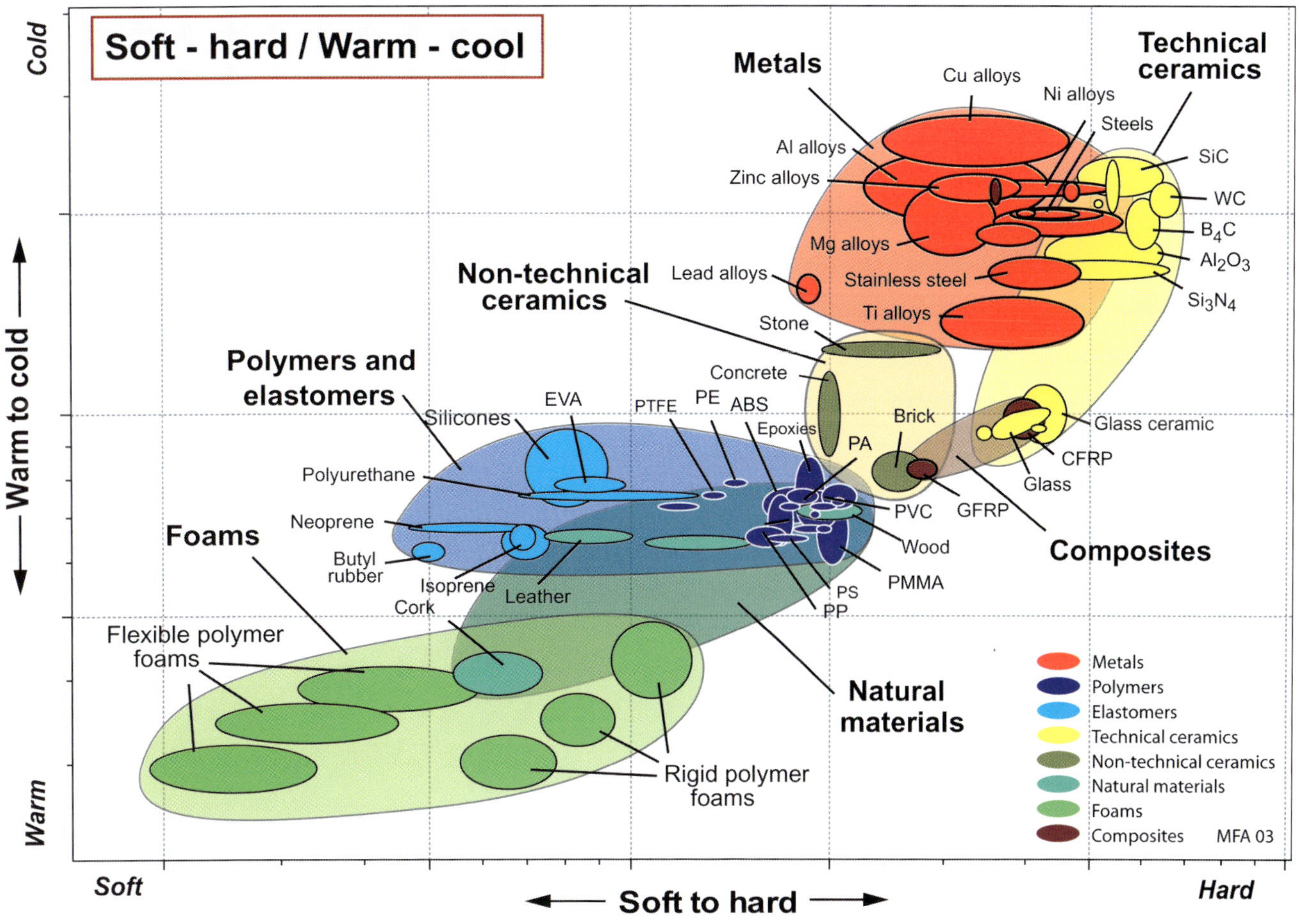

Figure 14 Tactile qualities of materials

Foams and many natural materials are soft and warm; metals, ceramics and glasses are hard and cold. Polymers lie in-between. (Abbreviations and acronyms are explained in Appendix A.)

A material feels 'cold' to the touch if it conducts heat away from the finger quickly; it is 'warm' if it does not. This has something to do with the technical property 'thermal conductivity' but there is more to it than that – it depends also on another property called 'specific heat'. A measure of the perceived coldness or warmth of a material is shown as the other axis of Figure 14, which clearly displays the tactile properties of materials. Polymer foams and low-density woods are warm and soft; so are balsa and cork. Ceramics and metals are cold and hard; so is glass. Polymers and composites lie in between.

Sight: transparency, colour, reflectivity

Metals are opaque. Most ceramics, because they are polycrystalline and the crystals scatter light, are either opaque or translucent. Glasses, and single crystals of some ceramics, are transparent. Polymers have the greatest diversity of optical transparency, ranging from completely transparent to completely opaque. Transparency is commonly described by a four-point scale that uses easily-understood everyday words: 'opaque', 'translucent', 'transparent', and 'water clear'. Figure 15 ranks the transparency of common materials. In order to spread the data in a useful way, it is plotted against cost. The cheapest materials offering optical-quality transparency ('water clarity') are glass, PS, PET, and PMMA. Epoxies can be transparent but not with water clarity. Nylons are, at best, translucent. All metals, most ceramics and all carbon-filled or reinforced polymers are opaque.

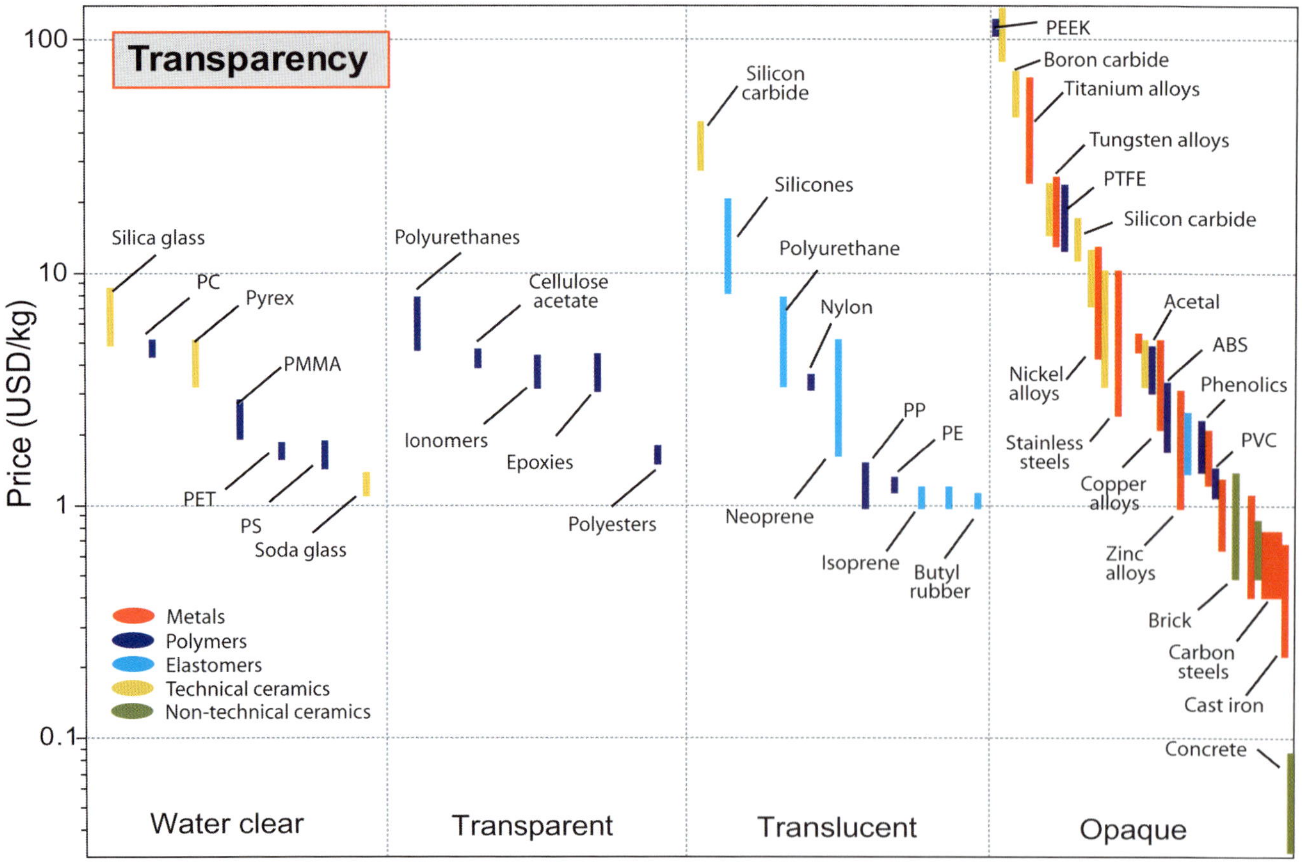

Figure 15 The transparency of common materials

Here transparency is ranked on a four-point scale, from water clear to opaque. Water-clear materials are used for windows, display cases and lenses. Transparent and translucent materials transmit light but diffuse it in doing so. Opaque materials absorb light. (Abbreviations and acronyms are explained in Appendix A.)

Colour can be quantified by analysing spectra, but this – from a design standpoint – doesn't help much. A more effective method is one of colour matching, using colour charts such as those provided by Pantone (see www.pantone.com); once a match is found it can be described by the code that each colour is assigned. Finally there is reflectivity, an attribute that depends partly on the material and partly on the state of its surface. Like transparency, it is commonly described by a ranking: dead matt, eggshell, semi-gloss, gloss, mirror.

Hearing: pitch and brightness

The frequency of sound (pitch) emitted when an object is struck relates to its material properties. A measure of this pitch is used as one axis of Figure 16. Frequency is not the only aspect of acoustic response – the other has to do with damping. A highly damped material sounds dull and muffled; one with low damping rings. Acoustic brightness – the inverse of damping – is used as the other axis of Figure 16. It groups materials that have similar acoustic behaviour.

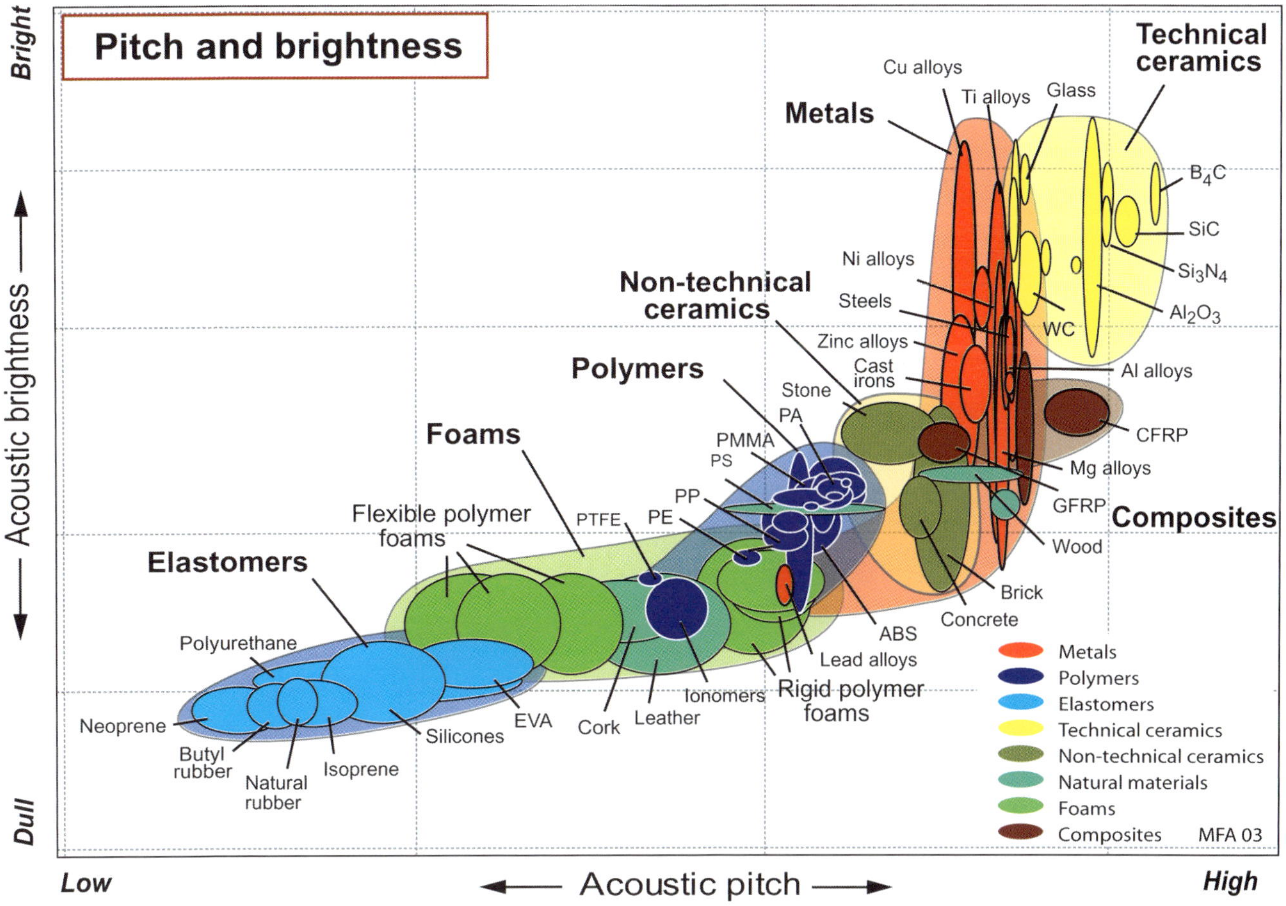

Figure 16 Acoustic properties of materials

A wine glass can be made to 'ring' because glass in an acoustically bright material with a high natural pitch; a plastic vessel of the same shape produces only a dull 'ping' because polymers are much less bright and – in the same shape – vibrate at a lower frequency. Materials at the top right of Figure 16 make good bells; those at the bottom left are good for damping sound.

Bronze, glass and steel ring when struck, and the sound they emit has – on a relative scale – a high pitch; they are used to make bells; alumina, on this ranking, has bell-like qualities. Rubber, foams and many polymers sound dull, and, relative to metals, they vibrate at low frequencies; they are used for sound damping. Lead, too, is dull and low-pitched; it is used to clad buildings for sound insulation.

Figures 14, 15 and 16 were produced using the CES software on the course DVD. Section 3 will introduce you to this software. The diagrams shown here use additional colour but are otherwise the direct output of the program. They show that each material class has a certain recognisable aesthetic character. Ceramics are hard, cold, high-pitched and acoustically bright. Metals, too, are relatively hard and cold but although some, like bronze, ring when struck, others – like lead – are dull. Polymers and foams are most nearly like natural materials – warm, soft, low-pitched and muffled, though some have outstanding optical clarity and almost all can be coloured.

These qualities of a material contribute to the product personality. The product acquires some of the attributes from the material, an effect that designers recognise and use when seeking to create a personality. A stainless steel facia, whether it be in a car or on a hi-fi system, has a different personality than one of polished wood or leather and that in part is because the product has acquired some of the aesthetic qualities of the material.

2.4.2 Materials and the mind: associations and perceptions

So a material certainly has aesthetic qualities – but can it be said to have a personality? At first sight, no – it only acquires one when used in a product. Like an actor, it can assume many different personalities, depending on the role it is asked to play. Wood in fine furniture suggests craftsmanship, but in a packing case, cheap utility. Glass in the lens of a camera has associations of precision engineering, but in a beer bottle, that of disposable packaging. Even gold, so often associated with wealth and power, has different associations when used in microcircuits: that of technical functionality.

But wait. The object in Figure 17 has its own sombre association. It appears to be made of polished hardwood – the traditional material for such things. If you had to choose one, you would probably not have any particular feelings about this one – it is a more or less typical example. But suppose I told you it was made of *plastic* – would you feel the same? Suddenly it becomes like a bin, a waste basket, inappropriate for its dignified purpose. Materials, it seems, *do* have personality.

Figure 17 A coffin; wood is perceived to be appropriate for its sombre, ceremonial function, plastic inappropriate

Expression through material

Think of wood. It is a natural material with a grain that has a surface texture, pattern, colour and feel that other materials do not have. It is tactile – it is perceived as warmer than many other materials, and seemingly softer. It is associated with characteristic sounds as it is worked or used, such as the thwack of a ball striking a bat, and smells, like the resinous scent of freshly cut pine. It has a tradition; it carries associations of craftsmanship. No two pieces are exactly the same; the woodworker selects the piece on which he will work. Wood enhances value: the interior of cheap cars is plastic, that of expensive ones is burr walnut and calf leather. And it ages well, acquiring additional character with time. Frequently, objects made of wood are valued more highly when they are old than when they are new. There is more to this than just aesthetics; there are the makings of a personality, to be brought out by the designer.

And metals... Metals are cold, clean, precise. They ring when struck. They reflect light – particularly when polished. They are accepted and trusted: machined metal looks strong, its very nature suggests it has been *engineered*. They are associated with robustness, reliability, permanence. The strength of metals allows slender structures – the cathedral-like space of railway stations or the span of bridges. They can be worked into flowing forms like intricate lace or cast into solid shapes with elaborate, complex detail. The history of humanity and of metals is intertwined: the naming of historical eras 'bronze age' and 'iron age' tell you how important these metals were; and their qualities are so sharply defined that they have become ways of describing human qualities – an iron will, a silvery voice, a golden touch, a leaden look. And, like wood, metals can age well, acquiring a patina that makes them more attractive than when newly polished – think of the bronze of sculptures, the pewter of mugs, and the lead used in roofs.

Ceramics and glass? They have an exceptionally long tradition – think of Greek pottery and Roman glass. They accept almost any colour; this and their total resistance to scratching, abrasion, discoloration and corrosion gives them a certain immortality, threatened only by their brittleness. They are – or were – the materials of great craft-based industries: the glass of Venice, the porcelain of Meissen, the pottery of Wedgwood, valued at certain times more highly than silver. But at the same time they can be robust and functional – think of beer bottles. The transparency of glass gives it an ephemeral quality – sometimes you see it, sometimes you don't. It interacts with light, transmitting it, refracting it, reflecting it. And ceramics today have additional associations – those of advanced technology: kitchen stove-tops, high pressure/high temperature valves, space shuttle tiles... materials for extreme conditions.

And finally polymers. 'A cheap, plastic imitation', it used to be said – and that is a hard reputation to live down. It derives from the early use of plastics to simulate the colour and gloss of Japanese handmade pottery, much valued in Europe. Commodity polymers *are* cheap. They are easily coloured and moulded (that is why they are called 'plastic'), making imitation easy. Unlike ceramics, their gloss is easily

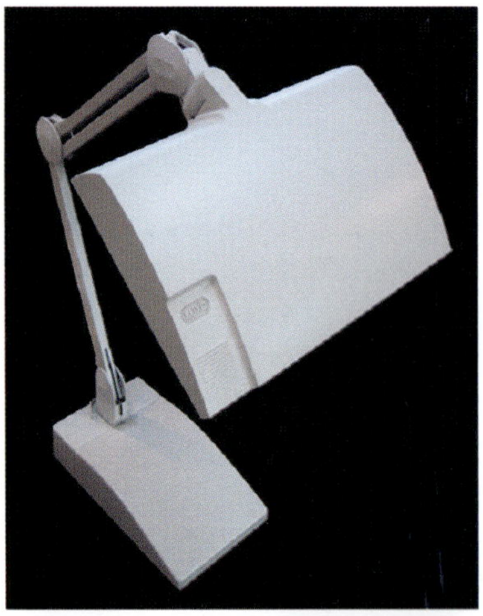

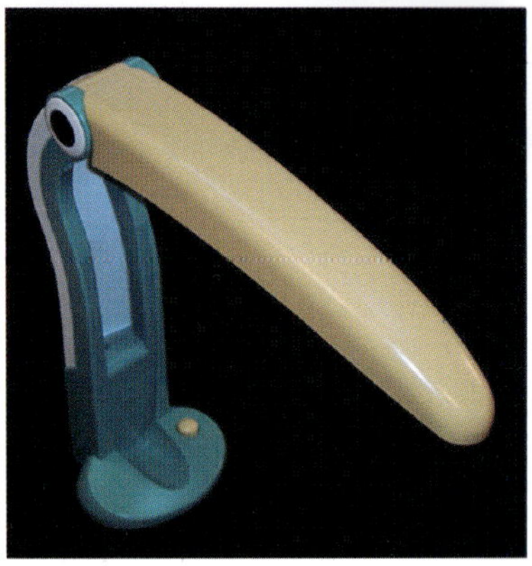

Figure 18 Lamps
Both lamps have the same technical rating, but differ completely in their personalities. Materials, processes, form, weight and colour have all contributed to the personality.

scratched, and their colours fade – they do not age gracefully. You can see where the reputation came from. But is it justified? No other class of material can take on as many characters as polymers: coloured, they look like ceramics; printed, they can look like wood or textile; metallized, they look exactly like metal. They can be as transparent as glass or as opaque as lead, as flexible as rubber or as stiff – when reinforced – as aluminum. Plastics emulate precious stones in jewellery, glass in drinking vessels and glazing, wood in counter tops, velvet and fur in clothing, even grass. But despite this chameleon-like behaviour they *do* have a certain personality: they feel warm – much warmer than metal or glass. They are adaptable – that is part of their special character; and they lend themselves, particularly, to brightly coloured, light-hearted, even humorous, design. But their very cheapness creates problems as well as benefits: our streets, countryside and rivers are littered with discarded plastic bags and packaging that decay only very slowly. Recently progress has been made in making such plastics easily recyclable or more biodegradable.

The ways in which materials, processes, usability and personality combine to create a product character tuned to the context or 'mood' are best illustrated by examples. Figure 18 shows the first. The upper lamp is designed for the office. It is angular, functional, creamy-grey, and, although this obviously can't be seen, it is heavy. Its form and colour echo those of computer consoles and keyboards, creating associations of contemporary office technology. Its form and weight transmit the ideas of stability, robustness, efficiency and fitness for task – but for tasks in the workplace, not in the bedroom. Materials and processes have been chosen to reinforce these associations and perceptions. The enamelled frame is pressed and folded sheet steel, the base-weight is cast iron, the reflector is stainless steel set in a high-impact ABS plastic enclosure.

The lower lamp in Figure 18 has the same technical rating of that on the left; the same functionality and usability. But the resemblance ends there. This product is not designed for the busy executive but for children (and adults who still enjoy being children), perhaps to be used in the playroom or bedroom. It has a contoured form, contrasting translucent colours, and it is very light. It is made of coloured acrylic in translucent and opaque grades, so that the outside of the lamp glows like a neon sign when lit. If this lamp were placed in a conventional office its personality would act to modify people's perceptions of that environment. Its form is partly derived from nature, partly from cartoons and comic strips, giving it a light-hearted character. I perceive it as playful, funny, cheerful and clever – but also as eccentric and easily damaged. You may perceive it in other ways – perception is a personal thing; it depends where you are coming from. Skilled designers manipulate perception to appeal to the user-group they wish to attract.

Figure 19 shows a second example. Here are two contrasting ways of presenting electronic home entertainment systems. In the lower picture: a music centre aimed at successful professionals with disposable income, comfortable with (or addicted to) advanced technology, for whom only the best is good enough. The linear form, the use of primitives (rectangles, circles, cylinders, cones) and the matt silver and black proclaim that this product has not just been *made*, it has been *Designed* (big D). The formal geometry and finish suggest precision instruments, telescopes, electron microscopes and the shapes resemble those of organ pipes (hence associations of music, of culture). The perception is that of quality cutting-edge technology, a symbol of discriminating taste. The form has much to do with these associations and perceptions, but so too do the materials: brushed aluminium, stainless steel and black enamel – these are not materials you choose for a cuddly toy.

Figure 19 Consumer electronics aimed at different consumer groups

The products in the lower image are aimed at a different consumer group than those in the upper. The personalities of each (meaning the combination of aesthetics, associations and perceptions) have been constructed to appeal to the target group. Materials play a central role here in creating personality.

Above this: electronics presented in another way. This is a company that has retained market share, even increased it, by *not* changing, at least as far as appearance is concerned (I had one 40 years ago that looked exactly like this). The context? Clearly, the home, perhaps aimed at consumers who are uncomfortable with modern technology (though the electronics in these radios is modern enough), or who simply feel that it clashes with the home environment. Each radio has a simple form, it is pastel coloured, it is soft and warm to touch. It is the materials that make the difference; these products are available in suede or leather in at least six colours. The combination of form and material create associations of comfortable furniture, leather purses and handbags, (hence, luxury, comfort, style), the past (hence, stability) and perceptions of solid craftsmanship, reliability, retro-appeal, traditional but durable design.

So there is a character hidden in a material even before it has been made into an recognisable form – a sort of embedded personality, a shy one, not always obvious, easily concealed or disguised, but one that, when appropriately manipulated, imparts its qualities to the design. It is for this reason that certain materials are so closely linked to certain design styles (Figure 20). A *style* is a short-hand for a manner of design with a shared set of aesthetics, associations and perceptions. We might identify a style with a certain period, but dates can only be approximate. Design styles do not switch on and off on specific dates, they emerge as a development of, or reaction to, earlier styles with which they often co-exist, and they merge into the styles that follow.

Arts and crafts: wood, textiles, natural materials

Art nouveau: wood, bronze, wrought iron

Modern and beyond: everything?

Art Deco: Bakelite, chromed steel, leather

Bauhaus: chromed steel, glass, plywood

Pop: plastic

Figure 20 Design styles and the materials they exploited to create product personality

The Early Industrial style (1800–1890) embraced the technologies of the industrial revolution, using cast iron, and steel, often elaborately decorated to give it a historical façade. The Arts and Crafts movement (1860–1910) rejected this, often choosing instead natural materials and fabrics to create products with the character of traditional hand-crafted quality. Art Nouveau (1890–1918), by contrast, exploited the fluid shapes and durability made possible by wrought iron and cast bronze, the warmth and textures of hard wood and the transparency of glass to create products of flowing, organic character. The Art Deco movement (1918–1935) extended the range of materials to include for the first time plastics (Bakelite and Catalin) allowing production both of luxury products for the rich *and* mass-produced products for a wider market. The simplicity and explicit character of Bauhaus designs (1919–1933) is most clearly expressed by the use of chromed steel tubing, glass and moulded plywood. Plastics first reach maturity in product design in the cheeky iconoclastic character of the Pop-Art style (1940–1960). Since then the range of materials has continued to increase, but their role in helping to mould product character remains.

Figure 21 Joining as a means of expression: fuel cap and watch

The precision-machined stainless steel fuel cap (top) is attached by eight Allen screws despite the fact it carries no significant loads.

Expression through process

Creating form is one of the earliest means of human expression: carved stone and moulded pottery figures, beaten ornaments and cast jewellery predate any documented ability to write or draw, exemplifying shaping as a channel for self-expression. The processes used in product design today are evolutionary descendants of these prehistoric antecedents. Figures 18 and 19 showed ways in which form and materials have been chosen and shaped to create the personalities of products, each designed with a particular user-group in mind.

Joining, too, can be used expressively. It reaches an art form in book binding, in the dovetailing of woods, and in the decorative seaming of garments. In product design, too, joints can be used as a mode of expression. The fuel cap of a contemporary performance car shown in Figure 21 (top), machined from stainless steel and attached by eight Allen screws, is an expression of precision technology that implies the same about the rest of the car. The watch (below), intended for sports-diving, uses the same motif to suggest robustness. The prominent welds on the frame of the mountain bike of Figure 22, left, suggests a prototype in development – perhaps at the forefront of current technology. The sleeve joints of the town bike on the right suggest a strong, dependable but unremarkable product.

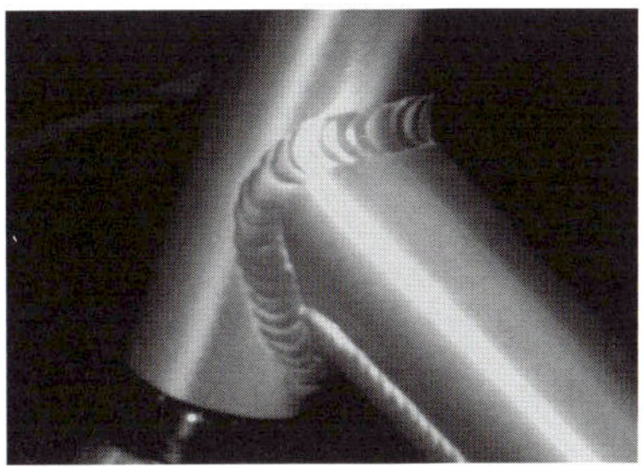

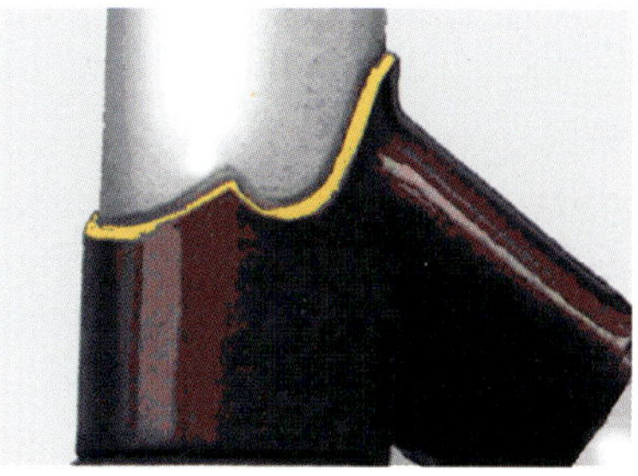

Figure 22 Joining as a means of expression: mountain bike and town bike

The bold prominent weld of the mountain bike on the left suggests prototyping – as if this bike was a latest development in a fast moving technology. The joint on the town bike, on the right, might suggest a strong but heavy product where styling and sturdiness are important.

Surface finish, too, carries messages. The late 20th and early 21st centuries are addicted to flawless perfection – surely a mistake! Surface perfection is violated by the slightest defect – it has no hope of ageing gracefully. Better, to make visual imperfection a part of the personality of the product – something that gives it individuality. It is this, in part, that makes natural materials – wood, leather and stone – attractive. Makers of earth-moving equipment have long delivered their products with a class A finish, the same as that required for a passenger car. And this, despite the fact that the first thing a purchaser does is to lower the thing into a hole full of mud to start digging. It is because the perfection of the finish expresses the perfection of the equipment as a whole; a poor finish implies, however mistakenly, poor quality throughout. Look again at the brushed aluminum and dyed leather of the products of Figure 19 and the way they create associations, the one of technical perfection the other of luxury handbags and luggage.

So surface processes can serve to attract, as with the digger. It can suggest, sometimes with the aim of deceiving; metallised plastic is an example. It can surprise, adding novelty – a jug kettle with a thermochromic surface coating that changes colour as the water heats up. It can entertain: holographic surface films can suggest something lurking inside the article to which it is applied. It can add function – non-slip coatings add an ergonomic function; contrasting colours identify different function-elements. And it can simply dress up the same product in different clothes, each to fit a different context (Figure 23).

Figure 23 Fuji Nexia Q1s: the same performer presented in four different costumes: blush, beach, cool, tech (clockwise from top left)

2.5 Using old materials in new ways

You don't need new materials to create exciting new designs. Innovation often means using old materials in new ways. Here are two examples.

Concrete

Figure 24 Glass-topped table with organic spiral shaped base made of reinforced concrete

Think, for a moment, about concrete. How do you perceive it? The pervasive view is that of a cold, grey material that is sterile, bland, utilitarian; the epitome of bare functionality. Its traditional and widespread use is in construction, from which most of its associations arise. It is the material of sewage pipes, of wartime pillboxes and bombproof bunkers, of the 'brutalist' architecture of public buildings and shopping centres of the 1960s. Not a personality you would welcome into your living room.

But think now of this. Concrete is a ceramic. It is hard and brittle, yet, before it sets, it can be cast (like a metal) or, moulded (like a polymer). Its texture and properties are easily adapted by varying the choice and the relative mix of cement and filler – fine sand, or gravel, or coarse aggregate. It is readily available, can be coloured, and it is very cheap. If it can be shaped so easily, could it not be used in more imaginative ways – ways that created new perceptions, acceptable in a domestic setting? The challenge is not just to find new uses, but to do so in ways that escape its present associations, allowing it to be perceived in new ways.

Figure 24 shows a glass-topped table with a flowing, sensuously curved tubular base made by casting concrete in a PVC tube wrapped round a cylindrical former. The free-standing organic shape escapes the usual perception of concrete as a sterile, bland material. This is one, among many, submissions to the British Cement Association Design Competition of 2003. They include concrete furniture, table lights and jewellery – all examples of what can be done by thinking of a material in a new way.

Wood

Figure 25 ALPA 12 SWA camera, ALPA® of Switzerland

The use of wood in a precision instrument such as this is unexpected but it creates the perception that craftsmanship has gone into the design.

Wood – as we have said – is a material of furniture, of musical instruments. But it is not something you would expect to find in advanced scientific equipment or precision instruments. Yet the designer of the camera shown in Figure 25 has chosen wood as part of what is a professional-level piece of photographic kit. The result is eye-catching; it sets the camera apart from the large number of high-level cameras now on the market. Because of the variability of wood, no two instruments are identical. Its presence diminishes the hostility associated with mechanical and electronic equipment, makes it a little more human and – as in the polished wood dashboard and panelling of an expensive car – it adds an association of hand-craftsmanship.

2.6 Summary and conclusions

What do we learn from looking at materials and product design?

The element of satisfaction is central to contemporary product design. It is achieved through an integration of good technical design to provide functionality, proper consideration of the needs of the user in the design of the interface, and imaginative industrial design to create a product that will appeal to the consumers at whom it is aimed.

Materials play a central role in this. Functionality is dependent on the choice of appropriate material and process to meet the technical requirements of the design safely and economically. Usability depends on the visual and tactile properties of materials to convey information and respond to user actions. Above all, the aesthetics, associations and perceptions of the product are strongly influenced by the choice of the material and its processing, imbuing the product with a personality that, to a greater or lesser extent, reflects that of the material itself.

Consumers look for more than functionality in the products they purchase. In the sophisticated market places of developed nations, the 'consumer durable' is a thing of the past. The challenge for the designer no longer lies in meeting the functional requirements alone, but in doing so in a way that also satisfies the aesthetic and emotional needs. The product must carry the image and convey the meaning that the consumer seeks: timeless elegance, perhaps; or racy newness. One Japanese manufacturer goes so far as to say: '*Desire* replaces *need* as the engine of design'.

Not everyone, perhaps, would wish to accept that. So, I end with simpler words – the same ones with which I started. Good design works. Excellent design also gives pleasure. The imaginative use of materials provides it.

A list of optional further reading for this section is given at the back of this block.

Key points of Section 2

Good design works. Excellent design also gives pleasure. Humans have a well developed capacity for attempting to link perceptions (for example based on the sense of sight or touch) to associations which can often lie at the root of pleasure in a product.

Products display, to a greater or lesser extent, a 'personality' and materials are a key determinant of this personality. Creating a desirable personality in a product is often an important aspect of the work of industrial designers.

It is possible to classify and measure certain properties of materials such as hardness. These measures allow the comparison of materials, facilitating the specification of material for a particular component.

Processes, such as joining and surface finishing, can also display personality.

Some materials, when used in particular ways, present their properties very directly. Other materials hide their properties or mimic other materials.

Some materials, such as wood and bronze, have a very long history of use. Their properties have been used and explored over many thousands of years. Other materials, such as some composites, are just emerging and their properties are only just being defined and exploited.

(Learning outcomes 1.2, 2.1, 2.2)

3 Materials and processes

In his essay in Section 2 Mike Ashby presents a strong case for viewing materials and processes as key determinants in a product's form, identity and personality. He referred to materials and processes as the 'flesh and bones' of a product. Without these we have only a design concept, perhaps existing only as a sketch or CAD model.

Ashby covered considerable ground in his essay. He introduced you to various classes of materials such as metal and ceramics, he made reference to some important processes such as joining, and he included a few charts which allowed you to compare and contrast material characteristics. In this section we will explore these in more detail by using the CES Selector which appears on the course DVD.

In Block 1, you were introduced to a wide range of chairs and one of the most important ways of describing the chairs was to describe their materials. In many instances you were provided with information about how the materials in particular chairs were shaped and formed into parts and how these parts were assembled. In the subsequent blocks, materials, and to a lesser extent processes, perhaps provide the most visible characteristics of the products you have explored. I want to start this section by looking, as Ashby did, at this relationship between materials and processes.

3.1 The relationship between materials and processes

3.1.1 General

Take a moment to look around your own environment now. The character of the floor beneath you might come from the particular texture of the carpet or the colour and grain of wooden planks; the feel of the book you're holding is a characteristic of both the paper and the printing process. If you had to describe your writing pen or your coffee mug you would probably include many features that resulted directly from the materials they are made from or the processes involved in their manufacture. In fact, trying to describe any product, from the smallest item to a whole building, without making reference to its materials and processes is extremely difficult.

When people describe products, they often reveal connotations or associations arising perhaps from the colour or the form of the product. As Ashby pointed out, the same is true for materials. People might describe the feel of a woollen fabric as warm, or associate the materials used in a kitchen gadget with hygiene, safety or modernity. Many of our perceptions regarding products are associated with the particular materials they are made from. Since these materials have usually been through various processing stages then these too can often influence our perceptions of products.

All materials have properties. Many of the chairs illustrated in Block 1 have been successful because someone has been clever enough to harness particular material properties to do a particular job (or in some cases more than one job). But of course, along with advantageous properties come disadvantageous properties. A certain wood might provide excellent visual appeal but offer little strength when used for a chair frame; a particular metal might be ideal for a tool but fiendishly expensive or difficult to process; a certain polymer might give a perfect seal to a sandwich box but taint the contents with an odour. As you must have realised by now, along with all the other problems that must be dealt with in the development of products, the designer or design team have to wrestle with the wanted and unwanted properties of materials and the wanted and unwanted properties of processes. Notice I didn't say 'good' and 'bad' properties. A polymer that is soft and bendy may make it bad for one application but ideal for another!

3.1.2 Materials and processes in a simple product – the clothes peg

Illustrated in Figure 26(a) are a number of examples of clothes pegs. They are not necessarily examples of good design – they simply come from my bag of pegs at home. However, they are (or were) made commercially so it's quite possible you could find similar examples in your own home. I want to use these pegs to further illustrate the relationship between materials, processes and design.

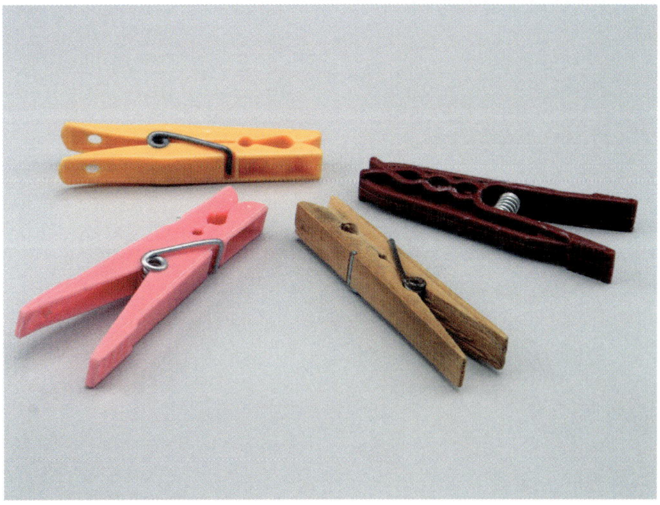

(a) Modern clothes pegs

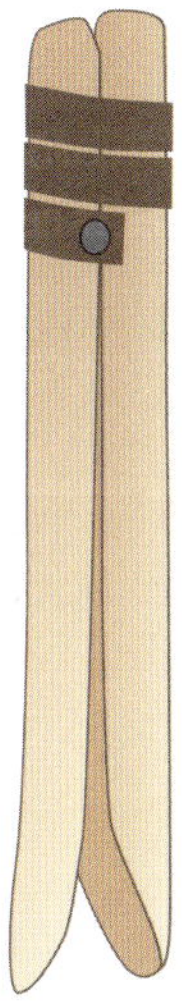

(b) Traditional wooden 'dolly' peg

Figure 26 Various examples of typical clothes pegs (UK)

Probably the most familiar clothes peg for most people in the UK will be the wooden peg that incorporates a metal spring. When you next have the chance, take one apart and examine it. The designs of these wooden pegs may differ slightly between manufacturers but the features and principles are the same. Notice that for any particular peg the two wooden components are identical – they are not made as a left hand part and a right hand part. This makes it easier for the manufacturer because the same process makes both halves, and there is no chance of getting the two components mixed up during assembly.

Notice also how the wood has been used. The grain runs down the length of the component, never across. This provides strength exactly where it's needed – where the forces are applied by the fingers, at the point where the peg pivots about the hinge, and where the forces are transferred from the two arms of the metal hinge to the wooden parts.

Today, if each peg was made individually, perhaps by carving and shaping each separate component by hand, clothes pegs would be prohibitively expensive. But pegs like these are made in their millions and are very cheap to buy.

SAQ 2

Traditional 'dolly' pegs (shown in Figure 26(b)) have been around for centuries. Speculate on the factors that made this type of clothes peg commercially viable when modern pegs must exploit mass production to be commercially viable.

Today the wooden components of modern clothes pegs are cut from large blocks of beech – a close grained wood that is good for machining. The grooves and tapers are cut by a high-speed router or machine saw, which passes across the grain of each block. Individual peg components are separated by simply cutting strips from these blocks. The spring steel is supplied as a roll of wire. Since the early days of the industrial revolution springs like these were coiled and bent using machine tools. Nowadays the process is fully automatic; even the assembly is automated resulting in a low-cost product of consistent quality.

At some point in time, probably in the years following the end of the Second World War, somebody said to themselves 'why go to the trouble of converting trees to planks of wood, seasoning them and then machining them into the components of clothes pegs when they could be moulded in plastic?' It was a good question. Moulding doesn't create the waste that machining does. OK, the mould tool might be expensive but, once the process is set up, it can produce millions of identical components. And so we saw, in the 1950s and 1960s, a wide range of home produced and imported polymer pegs made by injection moulding – a process of forcing hot molten polymer into a steel mould (Figure 27). If you look carefully at one of your own plastic pegs you might see a line around the edge of each part which marks where the two halves of its mould met.

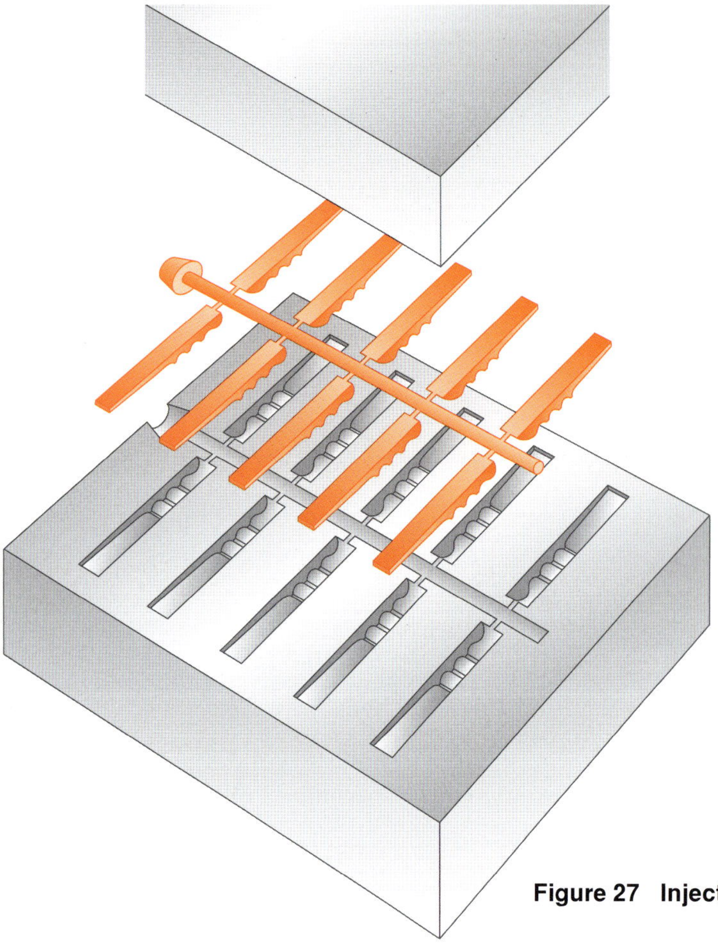

Figure 27 Injection mould for a basic polymer clothes peg

In the earliest types of polymer clothes pegs the polymer was used merely to emulate the shape of the wooden parts, that is to say, the form of the polymer peg was the same as the form of the wooden peg. The spring was identical. This is interesting because the form of the wooden peg derived from the particular qualities of timber, particularly the strengths and weaknesses arising from its grain structure, and the particular processing techniques used for wood (e.g. routing, sawing). If you look at some of the more modern designs for polymer clothes pegs you see a more creative process. You see pegs that acknowledge the qualities of polymers as a material and the qualities of polymer processing.

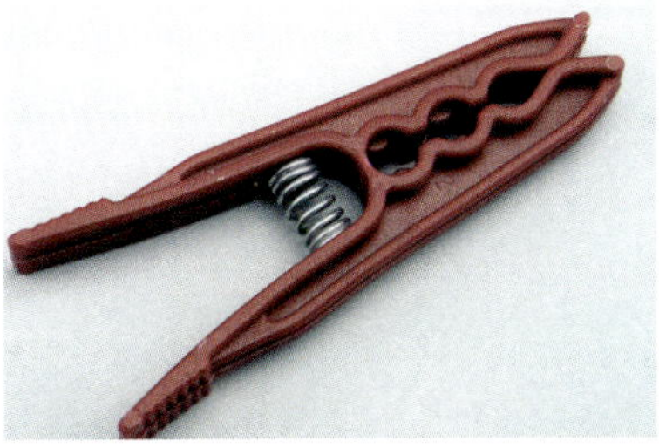

Figure 28 Two-part polymer peg

The clothes peg shown in Figure 28 possesses only two components (the polymer peg body and the spring) but this time the peg requires only one of each in its assembly (the previous examples required two of the polymer or wooden component plus one spring). The polymer part of the peg, that was previously two identical halves, is moulded in one piece. This peg is made from polypropylene which has very useful properties. It is not only strong but, if moulded thinly, it can fold and bend at particular points without breaking. It offers the possibility of incorporating an 'integral hinge'.

Figure 29 Close up of two-part polymer peg

Figure 30 Cross-sections of polymer (left) and wooden pegs (right)

If you look closely at Figure 28 you can seen that the peg pivots about a point near the centre of the moulding and because the material is flexible it will do this many thousands of times without breaking. Notice how the designer of this peg has achieved strength not simply by adding more material – the moulding has a complex section very different to the section of the wooden peg (see Figures 29 and 30). This 'I'-shaped cross section places material only where it is needed so the forces (for example, from our fingers and against the spring) are translated into an opening or gripping action. Let's look a little more closely at how this gripping force has been achieved.

A clothes peg needs to provide some grip to hold it, and the clothes, on the washing line. The two-part peg achieves its grip via a force applied by a spring placed between the inner faces approximately in the middle of the peg. This style of peg illustrates that cheaper and potentially better products can emerge if the designer or design team can integrate their skills with a broader knowledge. In Block 2 you were exposed to the importance of understanding the real nature of problems – particularly what the user wants from a product. In Block 3 you saw the importance of bringing creativity to designing. Here I highlight the importance of an understanding of material properties and an understanding of the processing techniques required in working these materials. But don't forget that designing is a process of being creative with materials *and* processes. Designing often involves the creative exploration and integration of the potentials of materials and processes as well as the application of existing knowledge. For some of the larger product companies this entails a close working relationship between material scientists, production engineers and design staff.

In the case of the polymer-and-steel-spring clothes peg, the company has reduced the number of components, reduced the amount of material needed, simplified assembly, exploited the potential of polypropylene and exploited a process very suitable for high volume production. It also offers the company an opportunity to market the pegs in a range of bright colours!

Consider this final example of a design for a clothes peg. Once again, it uses the potential of polypropylene but this time in a more sophisticated way. This peg consists of only *one* component. Figure 31 shows it in its open 'W' form, as it would emerge from the mould, and in its closed form. This peg also exploits an 'I' section to resist flexing and bending where this is not wanted. A small spur halfway along one side is simply snapped into a recess on the opposite side in order for the peg to pivot. Instead of a separate metal spring this peg has a plastic hinge moulded into it. The force that enables the peg to grip the washing line is the result of the polypropylene trying to return to its original open W shape.

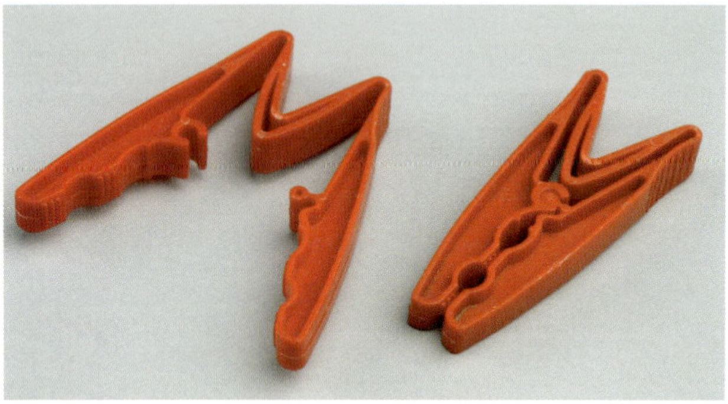

Figure 31 Two 'W' moulded pegs (open and assembled)

I judge this to be a very good product. The material is well suited to the user's needs in that it is unaffected by moisture or outside temperatures, it won't rust or otherwise stain clothes; it's cheap and light; it is strong, durable and easily cleaned. From the manufacturer's point of view it's a good design because each peg can be made for very low cost. It contains very little material and is made using an injection moulding process which enables many hundreds of thousands to be made from a mould. The peg has been designed so that the mould need only be a simple two-part mould. Also, there is only one component so sorting and storage costs are lower. Even the assembly is done by the purchaser (and thereby demonstrating its unique design to them!). It doesn't require instructions.

This would be a good time to refer to the *Block 5 Supplement* which presents a series of lessons on using the CES Selector. As well as the supplement you'll need your computer and the course DVD. Read Lesson 1 'Opening the CES Selector and navigating its content'. This will introduce you to the structure of this database and allow you to discover some of the relationships between materials and processes.

Lesson 1 and the other lessons each contain some computer-based exercises to provide you with an introduction to the database, its use in product analysis and its application in your own designing. If it's not convenient to view the DVD while you read this section of Block 5 I suggest you continue with the block to familiarise yourself with the material, and work through the lessons and the associated DVD exercises when it's convenient to you.

3.2 Classes of materials

The CES Selector materials and processes database on the course DVD presents six primary categories of materials and all the materials entries are located hierarchically under these six headings:

Ceramics

You are probably most familiar with ceramics as household decorative ornaments made from clay or perhaps china plates and cups. This group of materials includes many types of glass as well as minerals such as stone and hydrated ceramics such as concrete. It also includes silicon – the second most abundant element on earth, making up 26% of the earth's crust. In recent decades ceramics has become one of the fastest growing research areas for material scientists. A variety of new ceramic materials has been devised for a wide range of specialist and technical applications. Many of these materials will filter down into future consumer products, sports equipment and specialist applications as companies spot opportunities to exploit the particular properties of new ceramics. For example, boron carbide is a ceramic that offers low density and high hardness. It has been used recently to make the outer layer of bullet-proof body armour.

Composites

Composites were one of the great material developments of the 20th century. Like new ceramics, composites contain some of the most recently developed materials. You will see on the CES Selector that composites are available as either metal composites or polymer composites. The former are basically metals reinforced with ceramic particles. This can increase stiffness, strength and the working temperature of components without adding much weight – useful for aircraft components, for example, but now also being used in high performance sports equipment. Similarly, polymer composites such as carbon fibre reinforced composite are used to produce strong and light cycle frames for the top end of the market. Not all applications are so high-tech. Glass fibre reinforced polymers (GFRP) offer a cheap and low cost material for making a wide range of forms such as boat hulls and car body panels.

Foams

You may be sitting on some foam as you read this. There are two broad classes of polymer foam – flexible and rigid. Both are created by pouring or injecting a polymer, catalyst and foaming agent into a mould where they react, foam and cure. Flexible polymer foams can be soft and compliant and consequently they are often found in upholstered furniture and mattresses. Closed cell foams float on water whereas open cell foams absorb water (they are synthetic sponges). Rigid polymer foams offer a different range of properties. They are light and stiff and are often used as core material, e.g. sandwiched between cardboard for display panels or between aluminium to produce aircraft panels, or where thermal insulation or energy absorption is required. The rigid foam that you might be most familiar with is expanded polystyrene – found in many applications including packaging, flotation aids and safety helmets.

Metals

In his essay, Mike Ashby pointed out that the history of humanity and of metals is intertwined. The development and application of metals is indeed ancient but it hasn't stopped. Many new metal alloys have been developed in recent times. Metals are probably one of the easier groups of materials to identify – because they are generally cold to the touch, hard and reflective. Metals can be shaped in many ways including casting, rolling, stamping and machining which makes them very useful in industry. The CES Selector presents two categories of metal: ferrous alloys and non-ferrous alloys. Collectively their fields of application are immense, ranging from structures such as bridges and other architectural uses to foil and light bulb filaments.

Natural

Natural materials are those that have grown and been harvested. Some natural materials such as wood, bamboo and animal hide have been used since the earliest times to make shelters or clothing.

However, natural materials are seeing a resurgence in design, partly because of concerns with sustainability and partly because when combined with synthetics, natural materials can be given entirely new properties e.g. water resistance, vivid colour and longevity. There are a small number of natural materials presented on the CES Selector.

Polymers

Polymers present an increasingly large group of materials and they make up the biggest materials field on the CES Selector. Polymers can be divided into three sub-groups:

- elastomers;
- thermoplastics; and
- thermosets (thermosetting plastics).

Elastomers, as the name suggests, have elastic properties – they are often soft and pliable. This group includes natural and synthetic elastomers seen in such products as car tyres, wetsuits, rubber gloves and soles of training shoes. Thermoplastics soften when heated and harden again to their original state when cooled. This allows them to be moulded into complex shapes. You've already come across the ABS Lego block; other thermoplastic products include the non-stick surface on a pan (PTFE), plastic drinks bottles (PET), disposable transparent ball-point pens (polystyrene) and golf ball casings (ionomer). The thermoset polymers don't melt once they have been formed – they degrade. This group might be best illustrated by the epoxy resin 'Araldite', phenolics such as 'Bakelite' and 'Mylar' magnetic recording tape.

The examples given above are just a few of the many provided in the CES Selector database. If you haven't browsed the MaterialUniverse of the database yet return to Lesson 1 in Section 4 of the *Block 5 Supplement* to become acquainted with it.

A hierarchy such as that given in the CES Selector database is useful to designers. It enables them to structure the task of identifying materials that have been used in an existing product and it enables them to find a material that has the relevant properties for a particular design or component they are working on.

3.3 Materials properties

material property
A characteristic of a material which results from the materials' chemical composition and/or internal structure.

To make choices for design it is not enough that designers and engineers know the names and types of materials which are available; they need to understand the main properties of those materials. Every material has a set of material properties – strength, stiffness, hardness, electrical conductivity etc. – which depend on its chemical composition and its internal structure. It is usual in the teaching of materials science to explain how and why the properties of the main classes of materials differ in terms of their structures, starting at the atomic and molecular levels. For example, certain properties of metals can be explained in terms of their crystalline structure comprising a regular arrangement of atoms and free-moving electrons, while polymer properties depend on their structure of very long chain molecules entangled rather like cooked spaghetti.

In this course our aim is to allow you to develop a broad understanding of materials and their properties without going into the depth that students of materials science may require. To help you appreciate why materials properties are so important and what types of properties may need to be considered in design, try the following exercise.

Exercise 2 Materials properties

In choosing a material for making the following objects, what properties are likely to be essential or important? In each case try to suggest one or more suitable materials.

1 The blade of a screwdriver
2 The bodies of kettles
3 A car windscreen

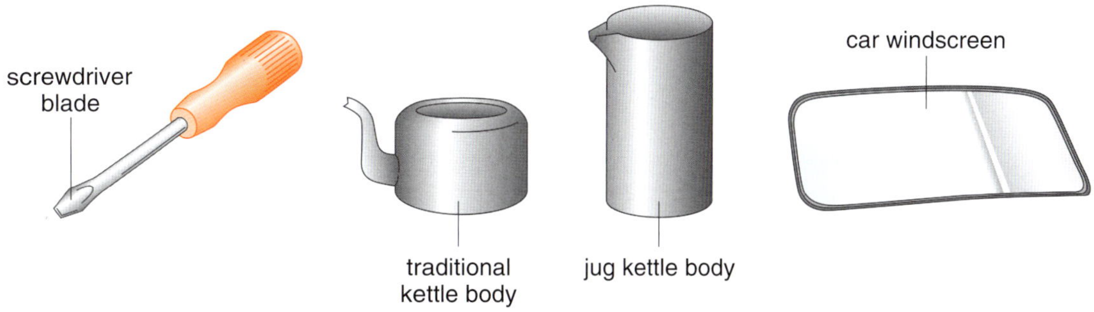

Figure 32 What properties are important in choosing suitable materials for these objects?

Spend a few minutes making notes on the required materials' properties and suitable materials for each object before looking at my suggestions overleaf. If you get stuck, or the terms are unfamiliar, don't worry – materials properties are considered in more detail after the discussion below. If you wish you can look up in the CES Selector on the DVD any of the materials I refer to below.

Discussion

1 The blade of a screwdriver must be made of a material that is *stiff* enough when made in the required shaft size so as not to twist or bend temporarily when turning a screw. It should be *strong* enough so as not to break or deform permanently if you twist it hard, as some poor quality screwdrivers do. The blade must also be *tough* so that it does not fracture suddenly when subjected to steady forces or impacts. A screwdriver is usually produced in large volumes to a low price, so the chosen blade material should be *easily processed* into the required form, must be low *cost* and should be readily *available* in suitable stock size for processing.

A material which satisfies these requirements is high-carbon steel, a metal. A ceramic such as glass would not be suitable because although it is stiff and hard, it is very *brittle* (i.e. it lacks toughness) and would break suddenly without bending or twisting first. Most polymers would not be sufficiently stiff or strong, while a composite such as carbon-fibre reinforced plastic would be unnecessarily expensive for this application.

2 The first requirement of the kettle body material is that it must be able to withstand boiling water without softening, melting or chemical change. It must be sufficiently *stiff* when made into a suitable shape so as not to distort significantly when in use. It must be *light* enough to enable the kettle to be easily lifted. It must not *corrode*, gradually distort, or otherwise deteriorate in use, and must not affect the taste or purity of the water. The material should not *cost* too much. It should be *available* in sufficient quantity, and must be capable of being economically *processed* into the required shapes. It must provide an attractive *appearance* and finish.

There are several materials which have suitable properties for making an electric kettle body, including stainless steel, aluminium, glazed earthenware, and some plastics. The material used will be selected on the basis of cost, appearance and finish, and ease of processing into the required thickness and shape. Both stainless steel and plastic jug kettles can easily be found side-by-side in high street retailers, revealing that they both offer commercially viable options. More traditional steel kettles that sit on top of a stove are still popular.

3 The material for a car windscreen must of course be optically *transparent* as well as *stiff* and *strong* enough in a suitable shape and thickness to satisfy its structural requirements (note how many mass produced cars nowadays exploit large transparent panels that possess complex compound curves). The chosen material for a windscreen must be *tough* enough to withstand minor impacts without fracture, and if it does fracture must not present a hazard to the car's occupants. It should be capable of being economically *processed* into shape without producing optical distortions.

A suitable material is toughened and laminated glass. Transparent plastics, such as polycarbonate, might offer some advantages but generally these scratch more easily and are likely to be more expensive. Marketing a car with a plastic windscreen may also be difficult – people perceive glass to be a higher quality material for this application.

From the above exercise you should have realised that there are two sets of overlapping materials properties which are important – the properties required in a final product or component, and the properties required for making that product or component. These two sets of properties often conflict. For example, thermoplastics can be readily processed into shape at relatively low temperatures, but this property means they generally make poor structural materials since under load they tend to 'creep', that is deform permanently over time.

It is very important therefore that designers have a good working knowledge of both the *performance properties* and the *processing properties* of materials. But choosing materials for their performance and processing properties is not enough. As the exercise should have made clear, the designer also has to consider the *economic properties* (price and availability) and *aesthetic properties* such as appearance and texture. This is represented by Figure 33.

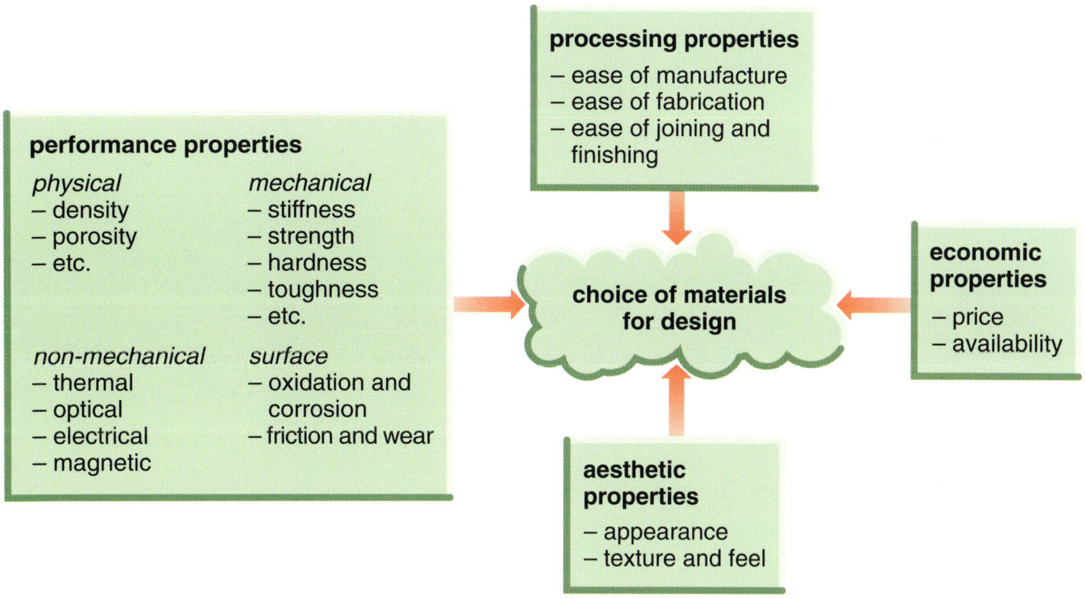

Figure 33 Different sets of materials properties and their effect on choice of materials for design

3.3.1 Materials and form

We began to look at some of the mechanical properties of materials in Block 4. You were introduced to terms such as strength, stiffness, toughness and hardness and we have revisited and added to these in the context of materials. Similarly you were introduced to shape and form and you began to see its influence on the structural performance of products and components.

The structural performance of a product or component depends not only on the mechanical properties of the material from which it is made but also on its geometrical form or shape. Thus, in the discussion of Exercise 2, I noted that the performance of the screwdriver blade, kettle body and windscreen depends on their shape and thickness as well as the properties of the chosen materials. The choice of material for a structure will therefore depend partly on the shapes in which it is supplied or can be processed.

Common experience tells us that one can make a structure stronger by increasing the thickness of the material. However, the design of many common structures is governed by stiffness; if the structure is stiff enough it is usually sufficiently strong. Stiffness of a structure depends on shape as well as the thickness and elasticity of the material. Think how much easier it is to bend a piece of copper wire or rod than a piece of copper tube or pipe. Or how much stiffer a sheet of paper becomes when it is rolled into a tube or folded into corrugations. This is because a hollow tube is stiffer in bending than a solid rod or sheet made from the same type and amount of material. Likewise a corrugated sheet, an 'I'-shaped cross-section beam or a box section is stiffer than a simple rectangular or flat section of the same weight and type of material. These principles are employed in the design of a 'monocoque' car body as shown in Figure 34. A monocoque car body is like a huge tube or box made from thin sheet panels welded together, thus exploiting the shape of the whole body shell to provide stiffness. The structure is further stiffened by cross-members and by shaping the panels into hollow box sections, curves, ridges and corrugations (Figure 34).

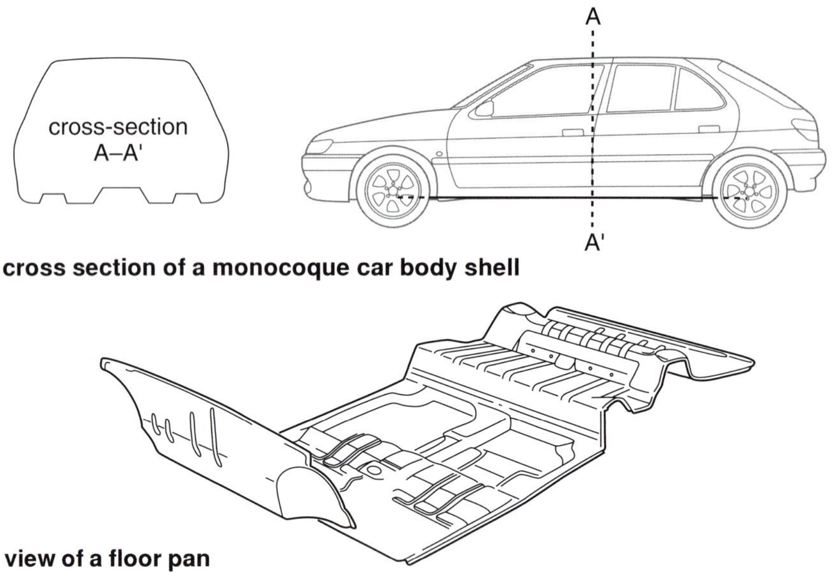

(a) cross section of a monocoque car body shell

(b) view of a floor pan

Figure 34 Example of the use of shape to provide stiffness in car bodies

3.3.2 Processing properties

Metals, ceramics, polymers, foams, composites and natural materials have combinations of properties which make them distinct. Yet in terms of the ability of different materials to perform a particular function, quite different materials may perform equally well, for example, alloy steel and glass reinforced polymers in car body panels. The choice between different materials may be better made in terms of their ease of processing. This choice is largely a question of balancing performance and processing – that is, reconciling the (often conflicting) properties required to achieve the various product functions with the properties required by the manufacturing process.

For some materials the conflict between performance and processing properties may be resolved by shaping the material while it is in an easily processed state and then *heat treating* it to produce the required performance properties. One of the reasons for the usefulness of carbon and alloy steels is that their strength, hardness and toughness can be altered by heat treatments. For example, a hacksaw blade is formed and the teeth cut and set with the alloy steel in a soft and ductile condition. The toothed edge is then heated in a gas flame and 'quenched' in cold water to produce hard, brittle teeth and finally 'tempered' by re-heating to 450 °C to reduce the brittleness while retaining useful hardness.

The properties of materials which are important for processing include ductility, malleability, viscosity and melting temperature. Such properties determine which processes are feasible and economic.

Ductility is the ease with which a material may be permanently deformed by tensile forces (e.g. pulling, bending) without breaking or cracking. *Malleability* measures how easily a material deforms by compressive forces (e.g. pressing, hammering) without rupture or cracking. Ductile materials (all of which are also malleable), such as aluminium, may be shaped in the solid state by processes such as pressing, rolling, drawing and extrusion through dies. Non-ductile materials, such as ceramics and wood, cannot be shaped using forming processes involving flow in the solid state.

Viscosity measures how easily the material flows in a liquid or semi-liquid state. Materials with low viscosities, such as molten metals, can flow under gravity and so can be cast; those with higher viscosities have to be forced into moulds, for example, the injection moulding of polymers.

The *melting temperature* of a material may be so high that processing from the liquid is impractical. If the material is also non-ductile, for example ceramics and tungsten metal, processing it in *powder* form may be the only feasible method, either by 'slip-casting' the powder in a liquid slurry or by processing dry powder or a paste and then heating ('sintering') to produce a solid. Powder processing is also used for forming a wide range of other materials.

Other properties can have a significant influence on processes, for example, *work hardening* – that is, the way in which metals become stronger, harder and less ductile under mechanical, cold working such as rolling or bending. Work hardening may provide useful improvements to the material; but sometimes because of work hardening the metal may have to be returned to a soft and ductile state for further shaping by a heat treatment called 'annealing'.

In addition to their knowledge of the processing properties of materials, engineers have developed a number of specific measures of ease of processing such as the 'machinability' of metals – an index of speed of metal removal, and the 'mouldability' of polymers – an indicator of suitability for various moulding processes. A number of broad indicators of material properties are included in the materials and processes database of the CES Selector.

Exercise 3 A feel for materials

Imagine you had a thin sheet of material from each of the main classes of material (metals, foams, ceramics, polymers, composites and natural materials). Imagine also that you were asked to form this into a curved shape by pressing each sheet into a mould (rather like a steel body panel for a car is made). What would the pressing activity reveal about the properties of each material?

Discussion

Let's think of the most unlikely candidates first, as their failure will reveal more about material properties.

Ceramics

A tile of ceramic material is non-ductile and brittle and would fracture without significant deformation. Also, its high melting temperature would preclude pressing it when soft. However, if we used a tile of soft clay, before it had dried and been fired, we could achieve a very accurate impression of the mould.

At room temperature glass is also non-ductile and very brittle. A cold sheet of glass would therefore crack and fragment when pressed. Glass when heated becomes less viscous and can be pressed into curved forms when soft.

Polymers

A sheet of polymer of the kind used to make the casing of a television set or a hair dryer would distort somewhat during bending perhaps becoming thin in places, before perhaps cracking and fragmenting. It would need to be heated to soften it before it could be pressed to shape.

A sheet of rubber would be undamaged because of its high elasticity (low stiffness). However, the material would return to its original shape – rather negating the point of pressing it.

Natural

A thin sheet of wood may bend easily but as we have seen in the discussion of Thonet's chairs (and as we saw in Block 1 in the discussion of laminating) wood does not retain the shape easily. If it is not pressed to the point where the grain splits it immediately springs back once pressure is removed. We have seen that steaming and binding to a former or glueing several sheets and then moulding is a good way to get wood to hold a curved form.

Foams

Depending on which foam you were using for this test you might find that it deformed and then split (e.g. polystyrene foam) or, like rubber, it deformed easily but sprang back when pressure was removed. Some foams are dense and brittle and will tolerate little bending. If a curved foam form was required it would be sensible to apply the foam in a liquid state and allow it to expand and cure (harden/set) in the mould.

Composites

A composite material, say a form of glass-fibre reinforced plastic, can be pressed into the required shape. It would act in a similar way to the sheet of wood – it would bend but spring back. Some composites can be softened by heating to allow moulding but an alternative way to achieve a permanent form would be to apply the composite materials in their liquid and flexible form and allow to cure.

Metals

Apart from the raw clay (that would need firing), the most plausible way for achieving a curved form of the same shape as the mould is by using metal sheets. You might use steel, aluminium, copper or many others. Metals have the right mix of strength, toughness and ductility.

Now would be a good time to return to the *Block 5 Supplement* and the course DVD to follow Lesson 2 for the CES Selector.

3.4 Materials selection

So far we have discussed materials mainly in terms of the six broad categories that are presented in the *MaterialUniverse* of the CES Selector. The tasks in Lesson 1 involve the various sub-categories and it is to these sub-categories that we now turn. It is not necessary for you to know all the names and properties of the different materials but it is very useful for you, or any designer or engineer, to have some understanding of the properties of the most common and useful materials within the sub-categories. This will enable you to make broad decisions on suitable materials and methods of processing for a particular application. It also allows you to better exploit information resources such as the CES Selector.

Metals may be divided into two categories:

- *ferrous* (those composed mainly of iron, e.g. the various types of steels); and
- *non-ferrous* (e.g. aluminium and copper).

Pure metals are generally too soft to be useful in engineering structures and so metal *alloys*, combinations of metal with other metal(s) or other elements, are employed.

Ceramics may be classified into five sub-categories:

- glass;
- glass bonded;
- hydrated;
- mineral;
- technical.

You will see from a short browse through the database that these sub-categories encompass domestic ceramics such as earthenware or china; engineering ceramics such as alumina, used for abrasives and cutting tools; various glasses such as soda-lime glass used for windows and some natural (mineral) ceramics.

Polymers are of two main types:

- *thermoplastics* which may be repeatedly softened by heating and then harden when cooled (e.g. the polyamides, a group containing various Nylon materials, used for gears and textile fibres; and polyvinyl chloride (PVC) used for many items such as drainpipes); and
- *thermosets* which can be softened only once for moulding and then harden irreversibly (e.g. epoxy resins used as powerful adhesives or urea-formaldehyde used for electrical fittings).

In the CES Selector, as in many other sources, lightly cross-linked thermosets (natural and synthetic rubbers) are classified separately as *elastomers*.

Composites are combinations of two or more materials with properties not achievable by the individual components. Broadly they can be defined as *metal composites* and *polymer composites* but other types exist. Composites can be classified by their physical composition. Many comprise reinforcing elements embedded in a matrix material, including fibre composites (e.g. carbon-fibre reinforced plastics) and particle composites (e.g. concrete). Notice that in the CES Selector, the entry for concrete appears under the category 'hydrated' within the 'ceramics' group, but the description opens with the line 'Concrete is a composite, and a complex one'. If designers are to find their way around such materials resources they need to have that broad understanding of materials' properties mentioned earlier.

The foams are also a type of composite in that they usually contain a matrix and an agent that produces a gas, which creates the foaming. Such has been the development in polymer science that foams now warrant their own material class. Broadly two sub-categories can be defined, *flexible polymer foams* such as those used in upholstered furniture, and *rigid polymer foams* such as the ubiquitous expanded polystyrene that is used for packaging of consumer products and for insulation in houses.

In some ways it would be logical to classify wood as a composite material, but a *natural* composite, because it consists of cellulose fibres and lignin. Nevertheless, it is probably less confusing to present it within its own group of natural materials including bamboo, cork and leather.

SAQ 3

We've come across plywood several times in the course. It's discussed at length in Block 1. Where would you place it in the hierarchical materials classification presented above?

Exercise 4 Matching properties to materials

Below are three sets of simple materials requirements which could have been extracted from different product design specifications.

- Identify into which of the main six material classes these material specifications fall.
- Try to select a material which broadly matches each specification. For each specification in turn, name both the broad category and the common, non-technical name for the appropriate material.
- Explain what aspect of the specification led you to choose one material rather than another.

You may be able to answer the above by just considering the requirements, but if you wish to you can refer to the CES Selector to assist you to address the tasks above.

The three materials requirements

1 High electrical conductivity, relatively low density, reasonable resistance to corrosion, processable as a liquid by simple equipment.

2 Electrically non-conducting, easily cast into complex shapes, must withstand deformation by up to 50% for long periods but quickly recover to original dimensions.

3 Relatively poor conductor of heat, usable at temperatures around 300 °C, optically transparent.

Discussion

In this exercise, it should be clear that we are dealing with the general categories of metals, polymers and ceramics. The appropriate materials which most closely match the simple specifications are aluminium alloys, rubber and glass, for the following reasons:

1 *Aluminium alloys*. High electrical conductivity is a feature of metals. So polymers and ceramics are ruled out by this criterion. Low density indicates that we can go straight to light alloys, ignoring ferrous metals. Aluminium alloy has a higher electrical conductivity and corrosion resistance than magnesium and is also cheaper – so this would be my choice.

2 *Rubber*. The requirement for electrical insulation rules out metals. Ceramics offer processability such as casting but only some polymers (e.g. partially crystalline thermoplastics and elastomers) have the potential to withstand large elastic deformations. It is only the latter group, containing the rubbers, which will recover their shape rapidly.

3 *Glass*. The need for low thermal conductivity rules out metals. The requirement for operation at 300 °C rules out polymers. The only group of ceramics to be optically transparent are the glasses – so a material can be chosen from these.

The experienced designer would know which broad categories of material match the requirements given in Exercise 4. Characteristic properties are inherent in each class of material, which would lead to initial selections in the manner given above. At later stages of design, when making more precise discriminations to gain further improvements in performance, a designer would go on to compare published information (such as that on the CES Selector), conduct tests and seek materials which most closely match the requirements of the product.

In Block 4 you were introduced to a variety of properties concerning, for example, technical function, aesthetic function, usability etc. You should now be viewing design activity – and particularly the phase where materials are selected – as a process concerned with establishing a balance or trade-off between multiple properties. Often the choice of material will be determined not only by performance and processing properties, such as those you have just considered, but also by aesthetic and economic considerations.

3.5 Recent innovations in materials

In his thorough and engaging book *Materials and Design* (Ashby and Johnson, 2002), Mike Ashby charts the way materials emerge and are exploited in products:

> New materials have their genesis in the laboratories of universities, governments and industry. The technologies they develop are spun-off, appearing first in 'demonstration' products, then absorbed into wider markets. As production volume increases the material price falls, and its increased familiarity to designers and consumers broadens the base of its use.
>
> (p. 157)

Ashby goes on to identify the attraction, and the frustrations, for the design community seeking to applying such new materials:

> To a designer, a new material offers both opportunities and risks. Opportunities derive from the new or improved technical or aesthetic behaviour that it offers. The risks – and they can be great – lie in the incomplete characterisation and the lack of design or manufacturing experience. It is usually the 'difficult' time-dependent properties – those relating to long-term integrity (resistance to corrosion, fatigue, creep and wear) – that are least well-documented and cause the most trouble. There is no historical body of design experience on which to draw, with the result that confidence in the new material is low and investment is seen as risky. Channels of communication are imperfect, making it difficult for designers to find the information they want. And, early in life, a new material is expensive – the development costs must be paid for – and availability can be restricted.
>
> (pp. 157–158)

But clearly advances in new products have, and continue to, come about through a close collaboration of materials scientists and those involved in design. It requires a willingness to integrate the often patchy nature of what is known about a new material with the creativity to explore its properties and to speculate on possible applications. It's another facet of the advantages of design team working that has emerged at several points in this course. Mike Ashby highlights a few recent examples where new materials have led directly to successful new products:

> Sports products have improved dramatically through the adoption of carbon fibre reinforced polymers, new elastomers, novel sandwich structures and – most recently – the use of titanium and amorphous metal alloys. Display technology, transformed once by the development of liquid crystal polymers, is about to be transformed again by LEPs – light-emitting polymer films. Advances in medical products have been made possible by new bio-compatible ceramics and polymers; and these, when successful, are very profitable. ... The successful introduction of polymer-matrix composites in aerospace was initially confined to non-critical components but now (2002) an all-composite wing for a large civil aircraft is a serious possibility.
>
> (p. 158)

Perhaps the most significant stimulus to the uptake of new materials by designers is the challenge, uncertainty and excitement that new materials offer. In professional design fields full of limitations and restrictions new materials perhaps offer the best opportunities to apply creativity.

Key points of Section 3

Materials and processes often present unwanted as well as wanted properties. These are not 'good' and 'bad' properties because their value depends on the context to which they are applied. Clever design can turn an unwanted property to advantage.

Some shapes suit particular materials; some materials demand particular shapes. The form of a product made in one material, say wood, may not be the best form if it is to be made in another material, say polymer.

Some processes offer opportunities to improve the design of a product from the users' and the manufacturers' point of view. The one-piece clothes peg offered a low-cost, and usable product with a long life. It also reduced component parts and removed an assembly operation. However, it probably required a higher initial investment by the manufacturer.

Six classes of materials, reflecting those in the CES Selector, have been discussed: ceramics, composites, foams, metals, natural materials and polymers. Each of these has been divided into sub-categories of material types (e.g. ferrous and non-ferrous metals). Each material within these sub-categories will possess properties including performance, processing, economic and aesthetic properties. Some of these properties may be common across the sub-category or even the class. Other properties may be unique to one particular material.

(Learning outcomes 1.2, 2.2, 3.1, 3.3)

Now would be a good time to return to the *Block 5 Supplement* and the course DVD to follow Lesson 3 for the CES Selector.

4 Processes: the manufacturing spectrum

In this section, I will ask you to think more closely about the processes of manufacture. We will consider the main classes of processing, the choice of a suitable process, and how this affects the choice of materials. As in Section 3, you will be pointed towards a few lessons in the *Block 5 Supplement* using the CES Selector and these will allow you to develop your understanding of processes and increase your ability to specify processes for a range of typical products and materials. Some worked examples in this section will allow you to see the decision-making process of others.

4.1 Types of production system

Before discussing specific processes it is useful to set them in the context of four basic types of production system. You came across some of these in the discussion of Thonet's furniture production in the opening section. The four basic systems are:

- *one-off* or bespoke production;
- *batch* production;
- *flow-line* production; and
- *continuous* production.

Different types of products and components are made using these different types of production system depending on the required *flexibility* of the system to produce different designs, and on the *production volume* required. Ships and buildings, for example, are usually produced as one-off products. Many industrial goods and engineering components are made in batches of various sizes using a mixture of general-purpose and dedicated equipment. Consumer products, such as cars and vacuum cleaners, are generally made on dedicated equipment by flow-line production. Continuous production in a dedicated plant is mainly used for bulk products, such as chemicals or fibres, which may not be identifiable as discrete items.

There are two reasons for introducing these basic production systems here. First, because references to them reoccur in the discussion of particular products and components throughout this section and, second, because certain manufacturing processes are best suited to particular production systems. Injection moulding of plastics, for example, involves relatively high levels of investment in equipment and dedicated tooling, and so is best used for large batch and flow-line production.

Mass production is a term that commonly means making similar objects in large numbers. It can be applied to any of the production systems listed, except of course, to one-off or bespoke production. There have been examples of mass production since the early years of the industrial revolution. But the term takes on a far more sophisticated meaning when we look at the system developed by Henry Ford in the early part of the 20th century. He devised a system for manufacturing and assembling standardised Model T cars economically, and in high volume. As the Thonet study illustrates, the basis of mass production was the complete interchangeability of parts and the simplicity of attaching them to each other. These developments enabled Ford to introduce the assembly line – marking a shift from volume production of cars using batch methods to mass production based on flow-line technology (Figure 35).

Figure 35 Model T cars at Ford's Trafford Park factory, Manchester, 1914

SAQ 4

What do you think might be the most appropriate type(s) of production system for making the following products?

- domestic refrigerator;
- racing bicycle;
- agricultural fertiliser;
- hi-fi amplifier;
- popular model of a family car;
- dining chairs sold by a major out-of-town warehouse chain;
- a unique sculpture commissioned from an artist.

4.2 Classes of processes

In the same way that engineering designers need a good knowledge of the range of materials, they equally need a good knowledge of the wide range of manufacturing processes. Once again the demand upon them is not to know everything about everything, but to know how the main processes relate, their strengths and limitations, where to go for information and how to proceed for detailed understanding. The starting point, as with materials, is a mental map of the widely established processes.

Lessons 1 and 2 of the CES Selector engaged you in an investigative process that brought both the MaterialUniverse and the ProcessUniverse into play. It's worth stopping to reflect on what the ProcessUniverse presents. It identifies three broad classes of process. The following paragraphs are edited versions of the statements that appear on the CES Selector.

If you haven't already done so, follow Lesson 1 of the CES Selector in the *Block 5 Supplement* which introduces the ProcessUniverse.

Joining

The manufacture of a product involves many steps, one of which is that of assembly – and it is one that adds cost. Two techniques contribute to economic assembly: design to minimise the number of components; and selecting the joining process that best suits the materials, joint geometry and the required performance of the joint during its life.

The CES Selector contains information for 19 of the most commonly used joining processes.

Shaping

Shaping processes give form to materials. The choice of process depends on: the material; the shape; the required size, precision and surface finish; and the associated cost, which depends most critically, on the number of components to be made – the batch size.

There are a number of broad families of shaping process. Some are economic, even when the batch size is small: vacuum forming of polymers and sand-casting of metals are examples. Others become economic only when a large number is made: injection moulding of polymers and die casting of metals, for instance. Variants of each process are adapted to be economic at a chosen range of batch size; and special processes have been developed that are viable for prototyping – where the batch size is one. (Section 5 looks at some rapid prototyping technologies).

The CES Selector contains information for 20 of the most commonly used shaping processes.

Surface treatment

Almost every component of a product has some sort of surface process applied to it. Surface processes enhance the thermal, fatigue, friction, wear, corrosion or aesthetic qualities of a surface, leaving the bulk properties unchanged. Economically they are of great importance, extending life, allowing materials to be used in increasingly harsh service conditions or imparting multi-functionality. For example, it's possible to create a corrosion-resistant surface on a tough, but chemically reactive substrate, or a thermally-insulating surface on a strong but metallic substrate. These surface treatments can create much of the visual and tactile impact of a product.

The CES Selector contains information for 23 of the most commonly used surface treatment processes.

Aim to complete all the lessons for the CES Selector in the *Block 5 Supplement* by the end of Section 4.

4.3 Choice of manufacturing process

4.3.1 Overview

Given the large number of shaping and joining processes, how do members of a design team decide which to use for making a particular product or component? As with the selection of materials there are two stages in process choice.

First it is necessary to identify a feasible process. Often what is feasible is determined by the expertise and equipment available in a particular factory, but assuming that a choice is possible there are two critical initial considerations:

- Can the process make the *shape* required?
- Is the process suited to the *material* to be used?

The first constraint is important because most processes are suited to producing some shapes and not others. Sand casting of metals for example, is well suited to producing complex shapes and sections at least 3 mm thick. It is not suited to making shapes of thin material; this is better done by pressing sheet metal or pressure die casting. Injection moulding of polymers can produce complex shapes, while compression moulding is best suited to simple shapes.

Secondly, most processes are suited to some materials and not others. For example, extrusion can be used to shape ductile metals and polymers, but rarely ceramics. Machining (e.g. milling), however, can be used to shape most materials given suitably hard and tough cutting tools.

These two constraints help to narrow down the choice of manufacturing process. But as you will see shortly, the choice is not one-way or linear. A designer may decide to change materials or modify shape to accommodate a preferred manufacturing process.

Exercise 5 Matching processes to products

Which of the processes outlined below would you choose to make the following objects? Match the products to the processes. You may use the CES Selector to assist you if you wish but, depending on the level of your current knowledge, you may prefer to attempt this unaided.

The selected products:

1 stainless steel sink;
2 cheap container for retailing milk;
3 panels of a domestic central heating radiator;
4 clear polymer egg boxes;
5 aluminium engine cylinder block;
6 thermoplastics casing for a television set;
7 turbine component made from engineering ceramics;
8 thermosetting plastics saucepan handle;
9 aluminium frame for a greenhouse;
10 synthetic rubber door seal of a refrigerator.

The selected processes:

(a) *Blow moulding*. Suitable for hollow shapes in glass and thermoplastics. This will produce a mould line (the line along which the two halves of the mould meet) all around the object. If required a handle can be moulded at the same time.

(b) *Pressing and sintering*. Suitable for most materials, including ceramics and very hard metals which cannot be shaped by other methods. Powdered material is pressed in a closed die to form the object which is then heated (sintered) at high temperature.

(c) *Forging (closed die)*. Suitable for ductile metals. Produces a much stronger product with a better surface finish than casting. 'Flash' (surplus metal edges) has to be removed by grinding.

(d) *Pressing from sheet*. Suitable for a range of ductile metals.

(e) *Sand casting*. Suitable for metal alloys that become liquid at the lower end of the melting point scale, e.g. aluminium alloys (around 660 °C). Useful where surface finish is not crucial. Good for large objects because the alternative of die casting in a metal mould would involve very costly tooling.

(f) *Injection moulding*. Suitable for thermoplastics, thermosets, rubbers and polymer composites. Gives dimensional accuracy, good surface finish and enables moulding of integral ribs for stiffness. Useful if form needs intricate internal structure for attaching other components.

(g) *Compression moulding*. Suitable for polymers, especially thermosets. Good for objects with thicker sections.

(h) *Vacuum forming*. Suitable for thermoplastics in thin sections. Best where tolerances are not critical. Commonly used for packaging. Considerably less costly tooling than injection moulding.

(i) *Extrusion*. Suitable for ductile metals, rubbers and plastics. Good for producing long thin shapes including those with complex cross-sections.

Discussion

1 stainless steel sink (d);
2 cheap container for retailing milk (a)
3 panels of a domestic central heating radiator (d);
4 clear polymer egg boxes (h);
5 aluminium engine cylinder block (e);
6 thermoplastics casing for a television set (f) (g);
7 turbine component made from engineering ceramics (b);
8 thermosetting plastics saucepan handle (g);
9 aluminium frame for a greenhouse (i);
10 synthetic rubber door seal of a refrigerator (i).

4.3.2 Choosing the optimum process

Often there will be more than one feasible process for making an object. If so, a designer or engineer will want to choose an optimum process from the alternatives. This will require some criteria for making the choice. There were some hints in Exercise 5 on the kinds of criteria that may be involved. For instance, tooling costs were mentioned in the choice between sand casting and die casting, and between vacuum forming and injection moulding.

Different manufacturers use different criteria to identify an optimum process, but typically most of the following would be applied:

- *cost*: the capital cost of new equipment, the cost of dedicated tools such as moulds, the energy costs, the labour costs in setting up and operating the process, and the assumed rate of depreciation for tools and equipment;
- *cycle time*: how long it takes to process one item (part, component or product);
- *product quality*: the standards required in terms of performance properties, surface finish and dimensional tolerances, and how reliably that quality can be maintained over time;
- *flexibility*: how easy it is to produce different designs on the same equipment;
- *materials utilisation*: the amount of waste material that is generated during processing.

The relative importance of the different criteria will vary according to the circumstances. For example, a highly flexible process such as machining, capable of producing different products and design variants on the same equipment (e.g. a lathe), is more relevant when making one-offs or small batches of an item than when making large numbers of identical objects. Different elements of cost will also depend crucially on the volume produced: it may be worth investing in new equipment or expensive tooling for mass produced items which would not be justified for a small production run. Shorter cycle times and better materials utilisation also become more important as production volumes increase.

This course doesn't expect you to be able to make process choices at this level of detail, but it is important to understand the factors involved.

The CES Selector offers three guidelines to the initial selection of process. These relate to the three classes of process:

- *Joining*: choose the material(s) to be joined and the desired joint geometry.
- *Shaping*: choose the material to be shaped and the desired shape class.
- *Surface treatment*: choose the function of the coating or treatment and the material to which it is to be applied.

It is important to remember that there are other, equally important, processes that change the properties of a material, either to make it easier to shape or to improve its performance in use. Some of the shaping processes also change material properties.

SAQ 5

Can you recall from Section 3 some examples of processes which alter materials properties?

4.3.3 Phases of manufacture

It is also important for designers to understand how the different processes link together in the manufacture of a product or component. During manufacture, material is usually first shaped crudely; then more exactly into finished parts or components; and finally assembled. So if we think of the processes in sequence we have three main manufacturing phases:

- primary processing: e.g. making the basic shape;
- secondary processing: e.g. finishing a part or component;
- assembly: combining parts or components into products.

Some products or components can be made essentially by a single process, for example injection moulding a plastic bowl. But most require a mix of processes. Primary shaping may be done by forming, casting or cutting processes (e.g. forging, moulding, milling). Secondary finishing normally involves cutting (e.g. drilling, grinding). Assembly requires the use of joining techniques (e.g. fastening, glueing).

4.3.4 Using a 'basic change list'

Block 3 introduced a number of techniques for creative thinking; decisions regarding materials, processes and manufacture often require creative thinking. The following list offers a few typical questions a designer or design team might ask if they had to generate creative ideas for improving a component or product. The answers to these questions may have many and varied implications, for example, in the materials from which a product is made, the processes used in its manufacture and the shape or finish of the item. This list takes the form of a 'basic change list' (Dagger, 1985). It offers a systematised version of the questions which an experienced designer would ask when examining a product or component that required change.

Basic change list

1 Is there any simpler way of achieving a particular outcome?
2 How do other industries do it?
3 Can it be eliminated?
4 Can any part or function be eliminated or reduced?
5 Can any, or all, of the parts or functions be replaced by other components?
6 Can parts or functions be separated, possibly giving more but simpler parts?
7 Can parts or functions be amalgamated?
8 Can it be 'bought-in' more cheaply?
9 Can it be made of standard parts?
10 Would it help to make it a standard part?
11 Can a new or different principle be used?
12 Can it be smaller?
13 Should it be larger?
14 Is it stronger than necessary?

For most companies there will already be a large investment in the manufacturing equipment to produce existing products. Very good reasons are necessary to move away from a well-understood material and a proven manufacturing process to something new requiring major investment and involving risk of failure. So a great deal of effort is placed upon trying to develop a component with minimum changes to existing processes. The most basic approach would be to consider improvements to the design which are possible without major changes to the manufacturing process or materials used. The most radical approach is to change everything – form, material and process.

The basic change list identifies elements of the knowledge and data upon which detailed design work rests. You should be able to identify the following as parts of a designer's competence:

- awareness of alternative basic design concepts;
- understanding of appropriate manufacturing processes;
- familiarity with known solutions;
- understanding of trade-offs, e.g. cost of machines versus component weight;
- knowledge of properties of materials which fall within an acceptable range;
- knowledge of properties of materials during processing;
- awareness of various types and sources of data, e.g. the CES Selector;
- appreciation of costs of materials, processes and finishes.

Designers and engineers involved in this type of activity need what has been described as a 'zoom lens' ability (Platts, 1990). They need to move between an approach which opens up possibilities and an approach which closes them down, from generating ideas to evaluating them, in the sort of iterative loops presented diagrammatically in Block 1. To think like a designer you need to move between exploratory techniques which redefine a problem, such as the basic change list, to evaluative techniques such as checking properties of machinability and wear.

This oscillation from overview to detail, from divergent to convergent thinking, from exploratory sketches to databases of information, is as important in detailed component design as it is in the early phases of concept design, which were discussed in Block 3. The oscillations can be both rapid and iterative. Although you may detect that there is a general sense of order in bringing forms and processes into focus, I don't mean to imply that there is one single pathway through the data. Designers are guided by a sense of direction and familiarity with both the physical and the conceptual. The terrain is difficult and that sense of orientation comes from experience in tackling many design problems.

4.4 Design for manufacture

4.4.1 Overview

The process of designing involves thinking ahead to how people are going to use a product, to its environmental impacts, and to how it will be manufactured. Throughout this course it has been emphasised that thinking ahead to manufacture is a characteristic of good designing and in this section I will examine the relation between designing and manufacture more closely. Design not only looks ahead to the particular methods of manufacture to be used – casting, injection moulding, machining or pressing, for example – but also to more general features of manufacture. These include the extent and type of automation, how the product is assembled, the way that a factory is organised, the scale of manufacture, the arrangements with suppliers, where manufacture takes place and even the overall strategy of the company. It's for this reason that design involves the most senior management in any company and not just designers and engineers.

It is useful to be a little more precise about the relationships between different areas of design for manufacture. First, there are the direct connections between design and manufacture:

- design for materials;
- design for process; and
- design for assembly.

Second, there are the connections between design and the organisation of manufacture. For example, products for a mass market will be designed so that they can be manufactured in flow-lines and designs for a family of similar products will share common components, sub-assemblies and 'platforms' so that manufacture of different variants can be coordinated.

Third, we can also identify more general connections between design and manufacturing strategy. For example, if a company's designs are to be sustainable then a strategy for sustainable manufacture is required across the range of the company's products (see Figure 36).

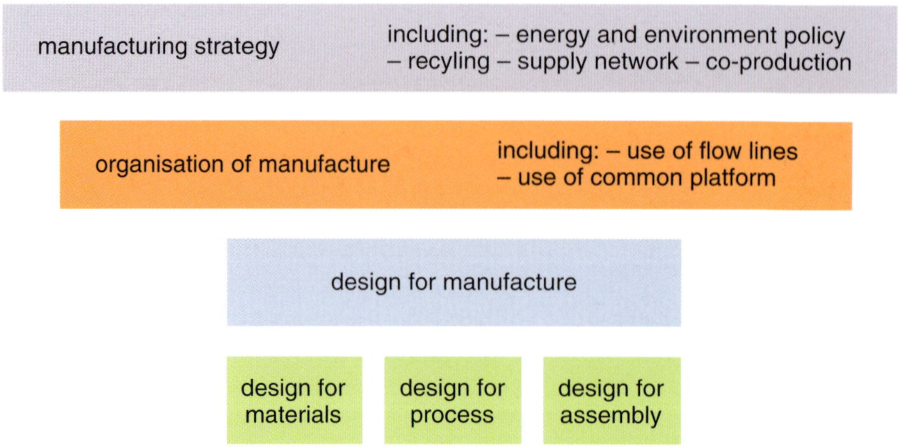

Figure 36 Diagrammatic representation of the applications for design within a manufacturing strategy

You have seen descriptions of several manufacturing processes. Design and manufacturing should not be viewed as separate activities. If they are then it is all too easy to create designs which are difficult or expensive to manufacture. The design process in such cases seems to proceed mainly by satisfying the functions required by users. The manufacturing processes, by contrast, can be a low priority, poorly understood and rarely acknowledged.

In these situations designs emerge in response to external factors. These designs are then passed to a manufacturer who is obliged to solve the problems of preparing the design for manufacture. This two-stage process of design followed by a separate manufacturing stage can give rise to problems in delivering a product to market. These include being expensive to manufacture, poor quality because of difficulties in controlling manufacture, or lengthy lead-times meaning that the product is delivered late to customers. A combination of all three problems is likely to result in a product failing in the market. Thinking ahead about manufacture at the design stage can avoid these problems.

Another problem with this 'over the wall' approach (so called because one aspect of a job is dealt with in isolation before being passed 'over' into another territory) is the possibility that the design generates deep antagonisms among management, design teams, manufacturing engineers and the shop floor.

Companies must consider manufacture during design if they are to deliver a product which meets user requirements of timeliness, cost and quality. These requirements are just as important as the functional requirements of the product we have discussed at length earlier in this course. Such consideration of manufacturing needs during design is sometimes called 'concurrent' or 'simultaneous' engineering in contrast to the 'over the wall' approach. More commonly we refer to this approach as *design for manufacture*.

4.4.2 Characteristics of modern manufacturing

As discussed above manufacturing is not just an add-on for a well designed product – it is essential to that product's success. A revolution in manufacturing has occurred since the 1960s. Companies, particularly in the Far East and Japan, started to understand how to combine design and manufacture to produce higher quality, more reliable products consistently. This understanding is now well developed and applied widely in companies around the world.

Design for manufacture contributed significantly to this revolution, but there were also considerable changes in manufacturing practices. However, before discussing these we need to understand a little more about what a 'high quality' product is. Essentially this is straightforward – the product should meet its various requirements. You have been well briefed on this throughout this course but reflect for a moment on the full extent of these requirements. A customer will value a product which is reliable, consistent, of appropriate cost and on time. The last might mean that a new product, promoted by the company and eagerly awaited by customers, appears on the market when promised and works properly.

How are higher quality products manufactured? The major insight here is that quality is improved if manufacturers can control their processes more effectively. If they can predict accurately how long operations will take, specify the variability in their products (e.g. tolerances) and increase the yields of 'fit for purpose' products (reduce waste) then they can 'tune' their manufacturing to meet competitive pressures. Manufacturers also want to reduce the overall times of product development and manufacture as well as the variabilities in these times.

Computer-controlled manufacturing

From the 1970s onwards, advanced technology and automation was one way to get control of processes. For example, computer controlled machines could manufacture more quickly and more consistently (with less variability). Today a vast number of types of product are manufactured using several computer-controlled machines integrated together. For example, the sub frames of a motor car are manufactured by processes which integrate computer-controlled cutting and pressing of sheet, robots for handling, and computer-controlled welding machines for fabrication. A drinks can is manufactured in a series of operations from cutting, pressing, coating, printing and shaping all closely integrated in a very fast automated process. This picture of integrated computer-controlled manufacture is seen throughout industry all over the world.

There are some advantages to using advanced technology and automation for manufacture in terms of speed and reduced variability but this comes at a 'price'. Such processes often involve expensive items of equipment requiring substantial investment. Designers need to think carefully about the limits imposed on their designs by the manufacturing capabilities available to the companies they are designing for. Designing outside these capabilities will increase costs if alternative manufacturers have to be engaged or cause delays if designs must be modified to fit manufacturing capabilities.

Quality control

Computer-controlled machines are expensive and will have a precisely limited (although extensive) range of possible operations. But advanced technology and automation are not the only methods to control variability and reduce production times. A key advance was the development of 'quality control' and 'quality assurance' functions. The former is based on better knowledge of what is happening in a manufacturing process using statistical methods – 'statistical process control' – in order to reduce defects to rare occurrences. The latter is based on conformance to established procedures which have been demonstrated to be effective such as the international standards for quality assurance (ISO 9001, ISO 9002, etc.). The attention to the detail of manufacturing methods required for controlling quality and programming computer-controlled machines means that manufacturing processes are explicitly specified. This in turn means that designers have access to information about manufacturing, and are not reliant on the experience of the manufacturing engineer and can codify manufacturing requirements into design rules. However, the recognition of the need for design for manufacture means that it is now common for a design team to include manufacturing specialists from the early stages of product development.

Cost reduction

A key cost in manufacture is the inventory of items held in stock before manufacture (input stock), the partially completed products during manufacture (work in progress) and the storage of finished products (output stock) before shipping to customers. These manufacturing costs can be substantial before receiving any income from finished products. Further, there are costs associated with handling all these items. How can appropriate design reduce all these costs? One way is to reduce the number of different parts used in manufacture. This might be accomplished by sharing standard parts across different products or reducing the numbers of parts in each product. Most car manufacturers, for example, will use various components across a complete range of cars.

Tolerances

As discussed in Section 1, all manufactured products will have some variability. Design takes account of this variability. For example, dimensional variability is specified in design by tolerances on the dimensions. The specification of tolerances is central to design for manufacture.

4.4.3 Design for manufacture – general principles

When designing a product is it useful to have in mind some general guidelines which are independent of any particular manufacturing process or method. These are mostly common sense and we have already met some of them earlier in this section, but they are worth listing as a checklist (see Box 2 overleaf). These are general principles and are not dependent on the particular characteristics of any one process. The next section takes a look at the design rules for some common processes.

4.5 Design for process

Each manufacturing process has its own design rules. In this section I've selected a handful of these in order to illustrate the kinds of rules a designer should use when designing a component to be manufactured by a particular manufacturing process. Of course, this is a huge subject with many different processes and I can only scratch the surface here. Of the processes selected, some should be familiar to you from the earlier sections of this block: forging, injection moulding, blow moulding, thermoforming and pressing from sheet.

> If you wish, you might want to use the CES Selector during this section to find out a little more about the various processes.

Box 2 General guidelines for design for manufacture

1. Try to minimise the numbers of parts and aim for simplicity in their shapes.
2. Where possible, use easily available materials and standard 'catalogue' components.
3. Standardise designs for similar products in a range using shared parts and sub-assemblies.
4. Define tolerances which are as large as possible whilst being consistent with product function. Unnecessarily small tolerances are very expensive because they require special tools and finishing.
5. Use easy-to-process materials.
6. Aim to design products using project teams containing both designers and manufacturing specialists.
7. Make components in single manufacturing operations where possible.
8. Design for the expected quantities required in the market – mass market products will be manufactured differently from one-off products.
9. Try to exploit the materials and the potentials of processes by which a product is to be manufactured.

Forging

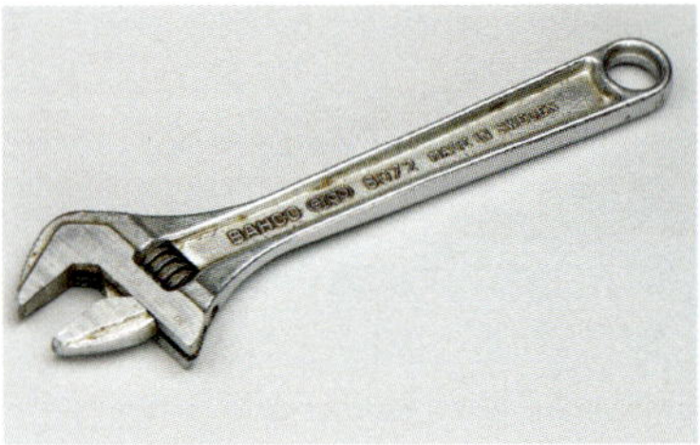

(a) Adjustable spanner

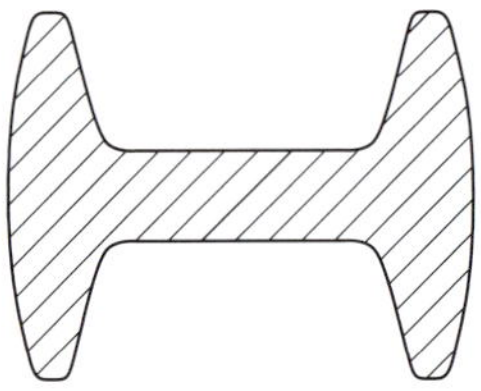

(b) a cross section of the handle

Figure 37

The adjustable spanner in Figure 37 is a product made from forged components. *Closed die forging* involves forcing a billet (a small bar) of hot metal between two dies. Usually the final component results from a sequence in which the hot metal is first forced into a rough shape by one set of dies and then transferred to another set of dies to refine the shape. Designers choose forged components when strength and toughness are required properties. As a consequence forging is used for components, which need to be strong but low in weight, such as hand tools or aircraft parts. The 'economic order quantities' (this term is explained below) for forgings are quite large because the dies and tools are expensive. For the handle of the spanner in Figure 37 the economic order quantities are about 3000 whilst for smaller forgings such as gears the economic order quantities are much larger. Here I should explain what 'economic order quantity' means. The cost of manufacturing a run of products is made up from two types of cost:

(a) *fixed costs* which are constant and independent of the number of products manufactured, e.g. marketing, factory space;

(b) *variable costs* which depend on the number of products in the run, e.g. raw materials, labour.

As the number of products in the run increases the total manufacturing cost per product goes down. The economic order quantity is the number of products in a run where the total cost per product matches what the customer is willing to pay. If a designer estimates a likely demand for a product below the economic order quantity for a particular chosen process then redesign may be necessary to reduce manufacturing costs perhaps by using a different process.

Forged components have special shapes. Importantly the shapes are tapered so that the forgings can be removed from the dies. The angle of taper or 'draft angle' for a steel forging should be about 5 degrees. On the spanner you see ribs on each side of the handle with a web in between. Ideally the depth of ribs and webs are roughly equal as in the spanner; again for easy removal from the dies. The edges of a forging are rounded – as in Figure 37(b) – to reduce possible sites for cracking and once again for easy removal of the component from the die. All edges of the spanner are rounded, except the surfaces to grip the nuts which are machined flat after forging. So we see that designing for forging puts many constraints on the shape but it does allow relatively small amounts of material to be used for a given strength.

Injection moulding

The Lego Duplo toy block in Figure 38 is an injection moulding of thermoplastic ABS. Heated thermoplastic is forced into a mould which is cooled. Economic order quantities are large, up to 100,000 and generally intricate and irregular shapes can be manufactured in this way. The walls of the parts are thin – about 1 mm for an ABS component like the Lego block. The point where the plastic enters the mould should not impair appearance. For the Lego block this is the round mark in the centre of the top of the block.

Figure 38

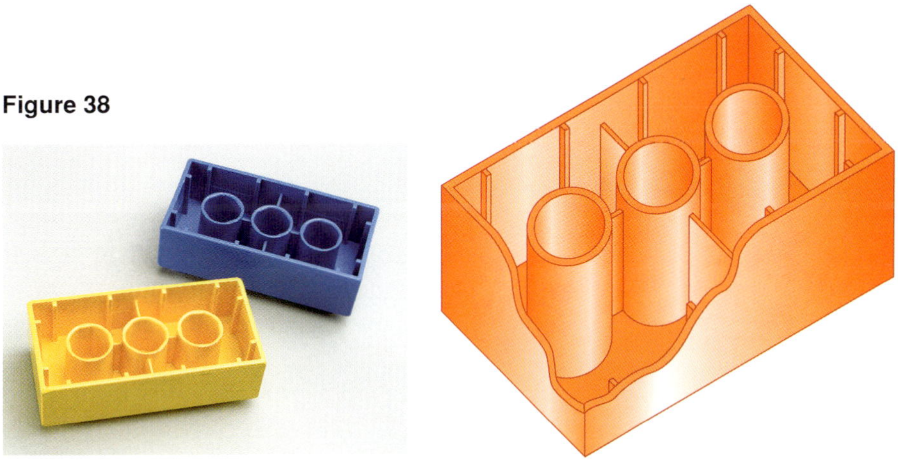

(a) Lego Duplo blocks

(b) Cutaway illustration revealing internal structure of ribs

On the underside of the block (Figure 38(b)) you see reinforcing ribs which should be thinner than the main walls (which they are for the Lego block) to prevent obvious 'sink marks' (indentations) on the opposite side of the wall. For injection moulded parts, draft angles can be negligible, and corners are rounded with small radii of about 0.5 mm which you can just see on the picture of the Lego block.

Moulds need to be tapered so that the component produced can be removed from the mould with ease. In this block we've called this taper the 'draft angle'. It is also known as the withdrawal angle, angle of draw, or just 'taper'.

Injection moulding can provide smooth shiny surfaces as in the Lego block, although most injection mouldings try to use matt finishes or gently curving surfaces to conceal sink marks. The Lego block with its glossy flat surfaces is relatively difficult to manufacture. Finally, all moulded parts show parting lines where the mould is opened. These parting lines are placed near sharp corners. For the Lego block they are placed right at the top and bottom of the flat sides. Designers will seek expert advice on how a design may have to be refined for the moulding process from specialist injection moulding manufacturers.

Blow moulding

Figure 39 Blow moulded detergent bottles with injection moulded caps

The detergent bottle in Figure 39 is a blow moulded component manufactured from PET. It has a complex shape with a complex injection moulded top and a built-in hinge for the cap. To produce the body, a thin bubble of PET is blown against the surface of a mould. The wall thickness varies across the product and can range from 0.4 mm to 3 mm (at the neck of the bottle for example); this variation is inherent in the process. The quantities produced for any given blow moulded component or product usually run into millions because there is little wear on the mould.

As with other moulding processes square shapes with no draft angle should be avoided as they are difficult to remove from the mould. Generally well rounded shapes should be used with large corner radii (> 5 mm). The shapes are somewhat constrained by the process of controlling wall thickness and mould cooling. The height of a bottle should be less than 15 times the diameter of the neck and the diameter (or width for a non-circular shape as in Figure 39) of the bottle should be less than about four times the diameter of the neck.

Thermoforming

Figure 40 Thermoformed tray for chocolates manufactured from polypropylene

Thermoforming heats a sheet of thermoplastic (e.g. ABS, PVC, polystyrene, polypropylene) and draws it onto the mould by vacuum or other means. Common applications are packaging for consumer products and confectionary (such as the formed tray for a box of chocolates in Figure 40, food containers or drinking cups) and larger household items like refrigerator linings or baths.

Design guides for thermoformed components are:

(a) only one side of the component has a good finish – the side next to the mould tends to mark and be less attractive;

(b) components should be shallow with width at least twice the depth;

(c) avoid sharp corners – these should be rounded with a radius at least twice wall thickness;

(d) this is a low accuracy process especially in terms of wall thickness;

(e) stiffness can easily be increased by corrugations;

(f) economic order quantities are small as tooling is usually low cost.

Pressing/punching from sheet

A panel used for the side of a steel filing cabinet is shown in Figure 41. The edges are bent to form channels which conceal sharp edges. Holes in the panels are for fixings. First blanks are cut from sheets with holes punched at the same time. The blanks are then bent by presses. Other parts of the cabinet are formed in the same way and welded to the main panel. There is a huge range of products from brackets to ring-pull drinks cans, made by pressing metal sheet using specialist machines and processes.

(a) Pressed panel in steel filing cabinet

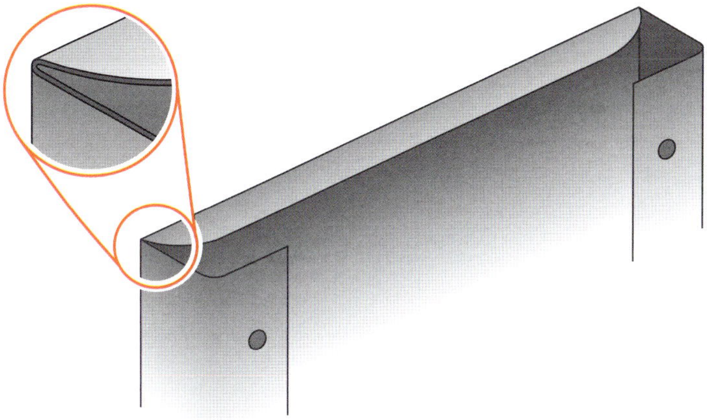

(b) Illustration of the folds in a pressed steel panel

Figure 41

Design considerations for parts made by punching and pressing include:

(a) wall thickness is uniform – drinks cans are about 0.3 mm thick and steel panels for furniture about 0.6 mm;

(b) the edges on cut and pressed pieces may be rough or sharp so avoid exposing them on the surface of a product;

(c) any channels should be at least twice as wide as they are deep;

(d) avoid shaped corners and narrow sections – tools tend to break prematurely;

(e) economic order quantities can be small (hundreds) but rise for specialist machines and tooling (hundreds of thousands in the case of drinks cans);

(f) design the blank shapes so that they nest closely together on the sheet from which they are cut to reduce waste;

(g) hole diameters should not be less than sheet thickness and multiple holes should allow at least twice the sheet thickness between them or from the edge of a sheet.

The examples of design for process considerations given above represent just a small number of the range of processes. However, you can observe some common points. Deep indentations and sharp corners are to be avoided if possible. The proportions between dimensions are critical. Complex, accurately made shapes with good surface finishes generally have large economic order quantities.

4.6 Design for assembly

4.6.1 General

In designing a product component the designer must also consider how all the components will be assembled into the finished product. This can be a complex and expensive stage of manufacture. If the designer can build into components features which make them easy to assemble – by machine or manually – then reductions can be made in assembly time and poor quality products arising from mistakes in assembly processes can be avoided. From our discussions of general characteristics of manufacturing you should be aware that lead time reductions and high quality are significant aspects of successful manufacturing. Design for assembly is therefore an area in which designers can have a significant impact.

Consider the toilet roll holders illustrated in Figure 42. It looks like it is more difficult, and will take longer, to replace the empty toilet roll in design (a) than design (b). We all have had experience of problems such as this, which are similar in their essentials to design assembly.

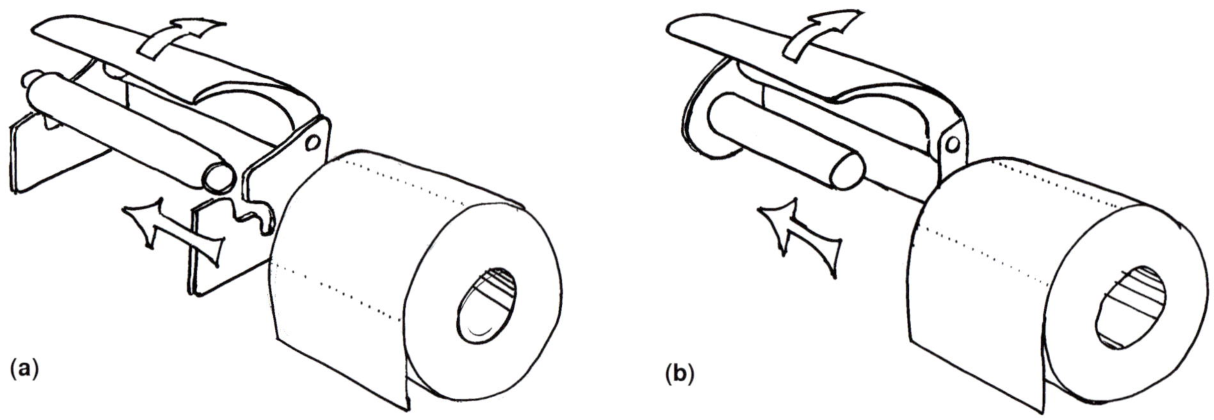

Figure 42 Two designs for toilet roll holders

Exercise 6 Stages in a process

Compare the process of replacing an empty toilet roll with a new roll in holder design (a) with the process in the improved design (b). What is the basis of that improvement?

Discussion

If we take the whole operation of changing the toilet roll as an *analogy of an assembly procedure*, then with design (a) we have a complicated assembly process. This entails several stages and handling of three parts in three directions in a complex series of movements – raising the lid; lifting out the cylinder; lowering the lid; removing the core of the old roll and discarding it; picking up a new roll and inserting it on the cylinder; raising the lid again; inserting the cylinder onto the holder; and finally lowering the lid.

Design (b) is an improvement, in which the assembly procedure is easier. Changing the roll involves fewer stages; handling of only two parts in two directions of movement; no movement of the cylinder and wide joining tolerances – raising the lid; removing the old roll and discarding it; picking up the new roll and inserting it on the cylinder; and lowering the lid.

This example has an echo of the irritation that ordinary artefacts can generate – as powerfully expressed by Donald Norman in Block 2. But here we can go beyond irritation and see that improvements are possible and that the improvements are based upon general propositions.

Such propositions include:

- minimise the number of components;
- reduce the number of operations;
- as far as possible, make assembly movements in the same alignment;
- make tolerances as wide as possible (in terms of assembling parts, not in manufacture itself);
- make the assembly self-jigging (that is, guiding surfaces and forms are built in to the form itself);

An example of aligning assembly movements is illustrated in Figure 43.

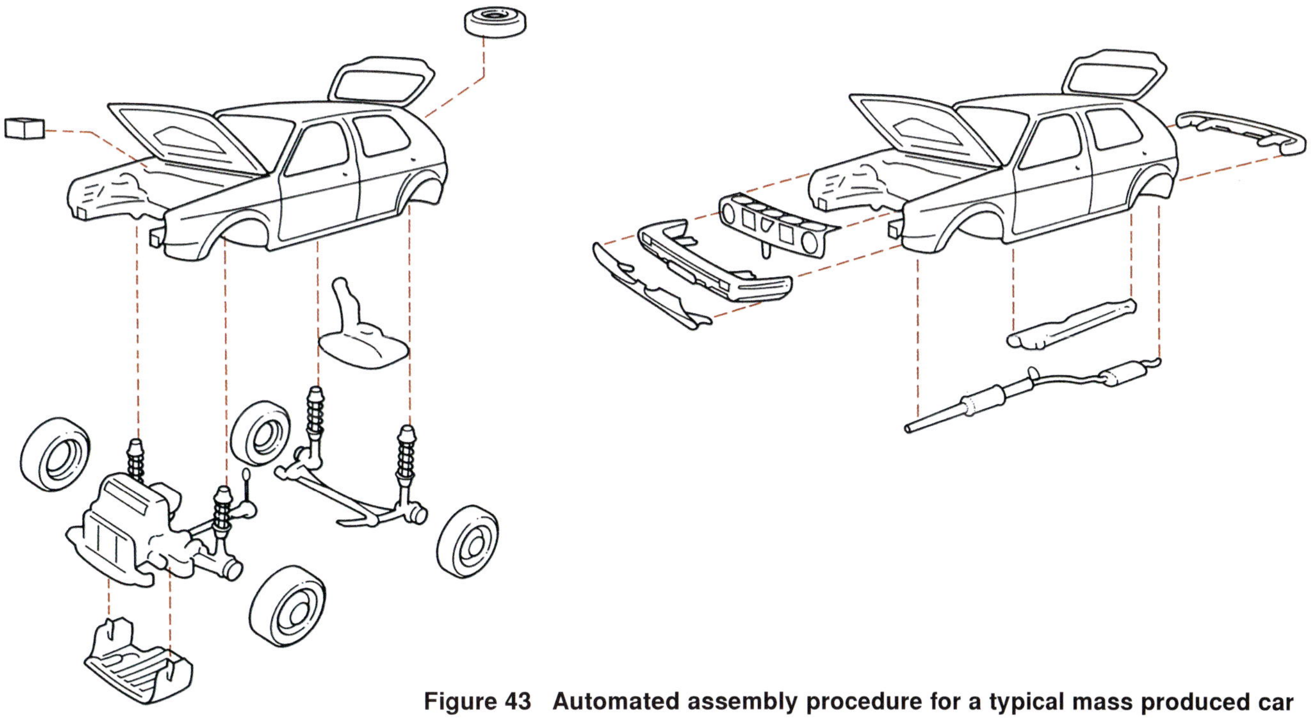

Figure 43 Automated assembly procedure for a typical mass produced car

Exercise 7 Minimising the number of components

If you were to follow the first rule in the list above – minimise the number of components – what changes would you suggest to the design of the toilet roll holder?

Discussion

It is conceivable that we could reduce the toilet roll holder to two, or perhaps even one, component as a horizontal wall mounted bracket or as a vertical cylinder and base standing on the floor. Variations of these ideas can be found in retail outlets.

4.6.2 The assembly process

The assembly process can be divided into four activities. The four are shown diagrammatically in Figure 44:

1 feeding to the site of assembly;
2 handling to the place of assembly;
3 locating in the assembly;
4 joining to other components by fasteners and/or more permanent fixing operations, for example, welding.

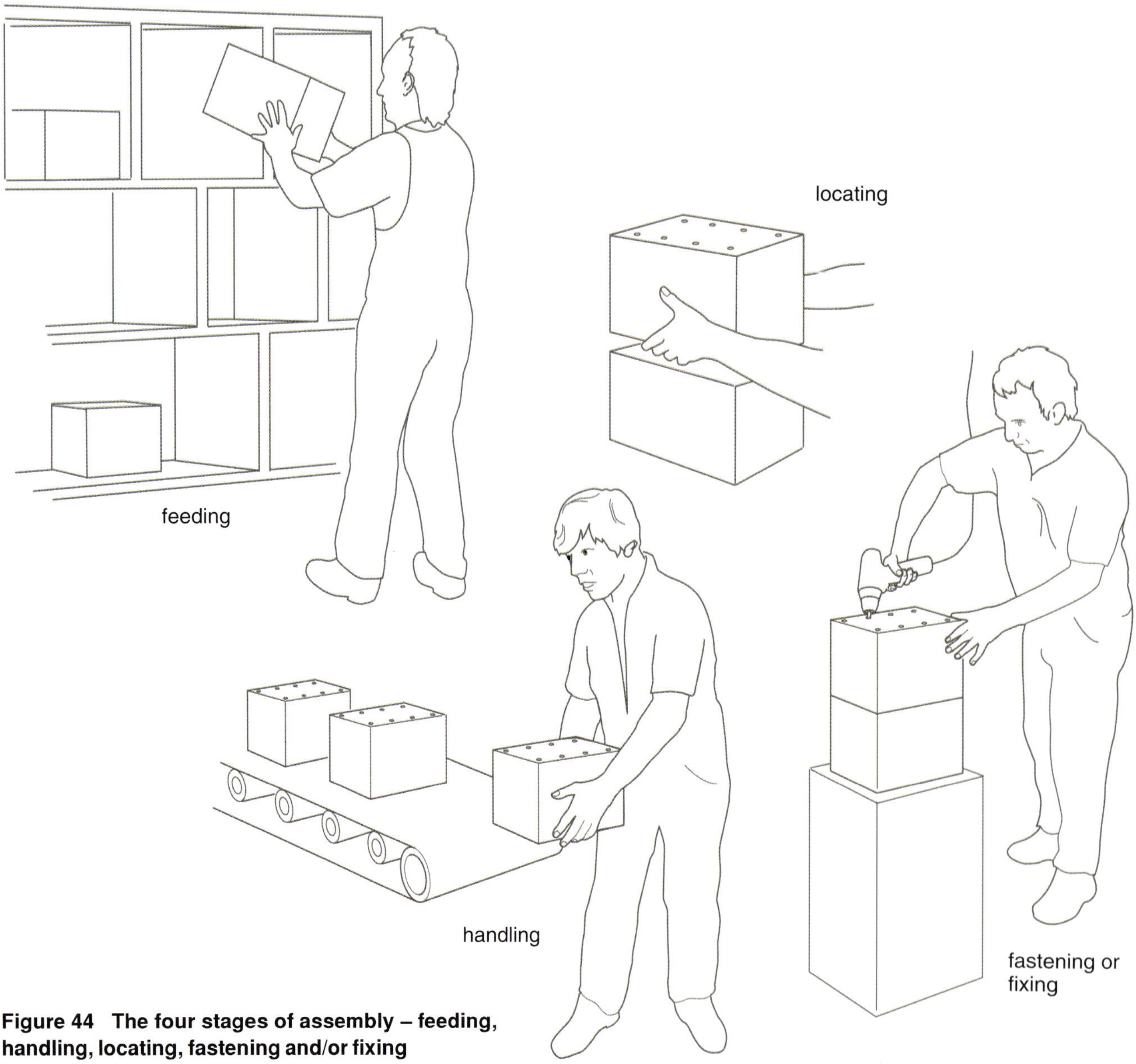

Figure 44 The four stages of assembly – feeding, handling, locating, fastening and/or fixing

Fastening may be done in several ways including screws, rivets, bolts, pins, staples, spring clips, adhesives or held together by tightness of fit or interlocking as in 'snap fits' (Figure 45). Fixing can be effected in a wide variety of ways including, welding, soldering, adhesives.

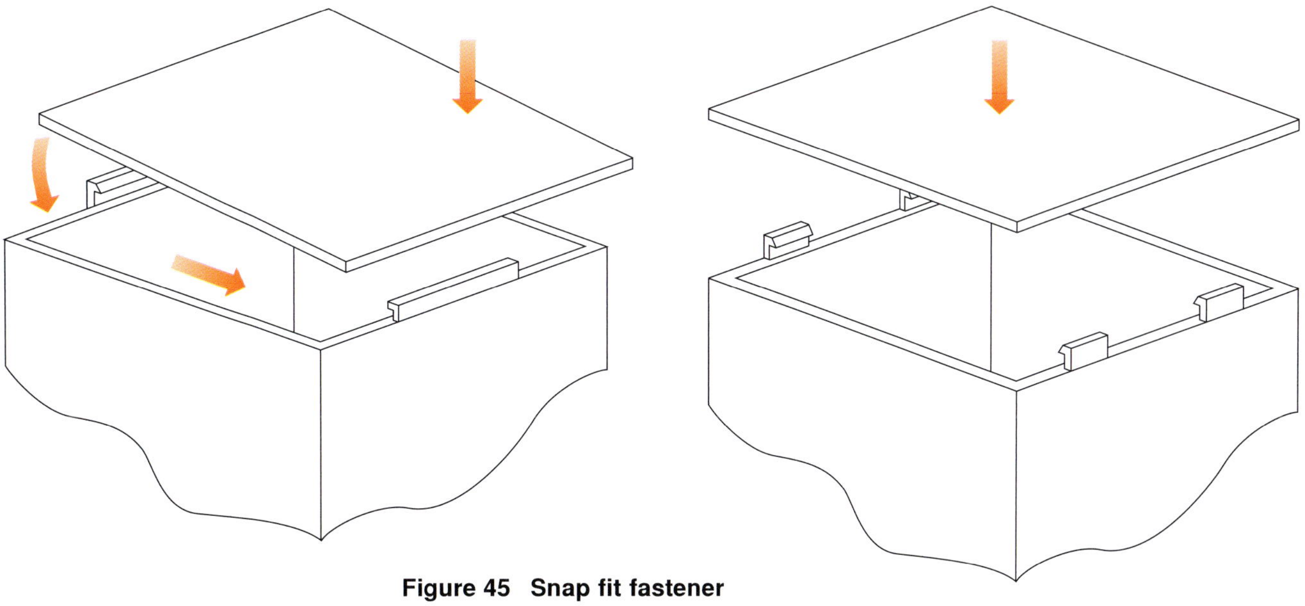

Figure 45 Snap fit fastener

In this illustration three directions of movement are replaced by a snap-fit lid requiring only downward movement.

Usually, the partially completed product passes from one assembly workstation to another. At each workstation several components are added to the product. In mass production a conveyor moves the partial assembly past several workstations with specific parts added at each workstation. This is an assembly line. Many different shapes of component, made from different materials will often be assembled in a single product. Some assembly lines are fully automated but many others still require manual work at various points in the assembly.

Assembly lines aim to be flexible within limits. Consumers expect products ranging from cameras to cars to be available in multi-feature options and as a basic option. The differences between products in a range should have as little impact on the assembly process as possible. This is done by:

1 making variation independent of assembly process operations – for example changing colours rather than shapes

2 enabling variation by including or excluding certain workstations from the path of a particular product. This means that the manufacturer can obtain variation by routing the product rather than changing assembly operations. This increases the requirements for scheduling the assembly – often handled by computer programs which control movement of products through the assembly line.

Higher levels of specialist assembly equipment and automation (in feeding, handling, locating and fastening/fixing) mean that the economic quantities of production for an assembly line between product changes become larger. The costs of changeover to a new product range can be reduced by using largely the same parts assembled on the same 'platform' but with significant differences in a limited number of parts. For example, computer companies aim for common arrangements of hardware between models, but with significant changes in computing power, usability and aesthetics (appearance and styling).

4.6.3 Design for assembly: general principles

Whatever the production quantity, designers can help to reduce costs and time of assembly. Designers should first of all try to visualise how parts will fit together. Computer aided design tools can help this as well as drawing and modelling.

The main principle in designing for assembly is to minimise the number of separate parts in a product.

Fewer parts mean less handling, fewer operations, less possibility of errors and higher quality. However, there is a limit to part reduction since beyond a certain point parts become too complex, heavy or large. With this in mind here are 17 general principles guiding design for assembly; Figure 46 illustrates some of them.

1 Combine parts.

   (a) If adjacent parts don't need to move, consider combining them.

   (b) If adjacent parts have to move, consider carefully how you might join them. For example, in some plastic parts, it's possible to incorporate hinges into the material, e.g. the top of the detergent bottle in Figure 39.

2 Consider snap fits and tabs instead of screws or bolts (as in Figure 45).

3 Group electrical and electronic components in one place, such as in mounting plates for several switches or printed circuit boards.

4 Eliminate parts which are not required in terms of the product's functionality.

5 Standardise components across a product range:

   (a) use standard screws, bolts and fasteners;

   b) reduce the number of different components across the range.

6 Use sub-assemblies – at final assembly a car is constructed from sub-assemblies such as engine, lighting units and instrument panels.

7 Design parts so that they cannot be inserted incorrectly.

8 Avoid the use of flexible parts, where possible, because handling can be difficult and they can tangle.

9 Use a layered assembly such that parts can be added from one direction.

10 Avoid the addition of several parts which must all be kept aligned whilst further assembly takes place.

11 Incorporate self-aligning connections between parts.

12 Use funnel shaped openings to holes and tapers on inserted parts.

13 Reduce handling of parts:

   (a) use shapes that can be fed by automatic machines;

   (b) increase symmetry so that parts fit in several orientations.

14 Ensure parts are easy to handle by people or machines.

15 Avoid tangling – e.g. use closed-ended springs.

16 Maximise tolerances to ease assembly operations.

17 Allow some flexibility or 'give' in components to ease fitting – the Lego Duplo block in Figure 38 uses the flexibility of the material to allow assembly of block after block.

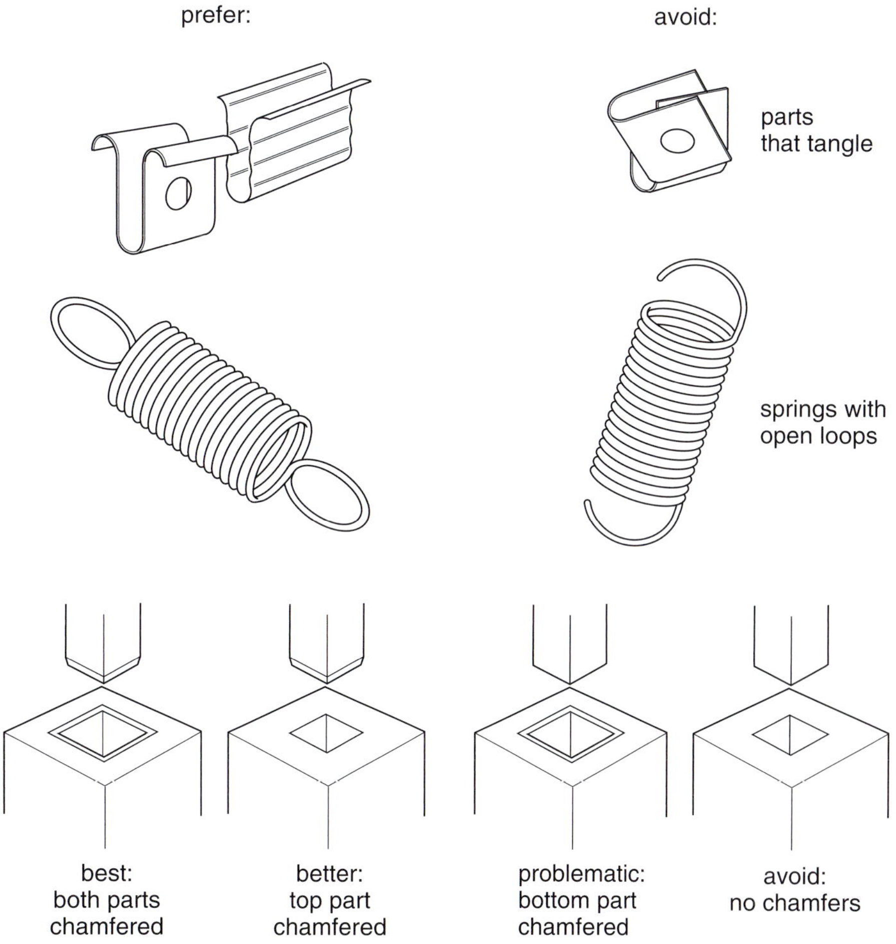

Figure 46 Illustrations of good design for assembly

This list of 17 general principles covers many of the design for assembly concerns of designers. However, there will be many other, more detailed principles which relate to particular manufacturing processes or assembly equipment. This list offers a good starting point, especially for mechanical assemblies.

4.7 Manufacturing organisation and strategy

4.7.1 Organising manufacturing: a history

The rapid development of computer controlled machines and better understanding of the importance of high quality and reduced product delivery times has meant that great attention has been given to the organisation of manufacture. Indeed this has been a 'second' industrial revolution. A little history might help.

Let us consider a crucial episode in the development of the young Toyota company. In 1950 Toyota was a small and relatively insignificant car manufacturer. Their annual output ran to no more than several hundred cars – tiny compared to Ford who, in their

Baton Rouge plant alone, produced over 7000 cars a day. The Toyota engineers realised they couldn't attain the same economies of scale as Ford in their manufacturing processes so sought a different approach. Take, for example, basic pressed components in steel. Large Western manufacturers like Ford were able to dedicate press machines to just one component and these machines ran, pressing out their particular component, for months or even years. The Toyota company set their sights on making their presses more flexible so they could make smaller numbers of a wider variety of parts. They found ways to make the change-over of dies (moulds) much quicker – by the late 1950s getting it down to minutes rather than the several hours typical in the Western corporations.

Toyota discovered, surprisingly, that there were great economies in their small batches of pressings. This was for two reasons. Firstly they had eliminated the storage and carrying costs of huge inventories of stocks. Secondly, making only a few parts before assembling them allowed errors in part design to show up almost immediately. Thus they had eliminated the waste of repairs and replacement. The repercussions were profound – the new designs for the press tools had reduced the need for specialist labour, the press operators could focus on product quality, the system facilitated instant feedback from assembly staff and the inventory was drastically reduced because the company only needed to hold sufficient parts for a few hours of car assembly.

This work on the *hardware* of production led Toyota to a complete revision of the *production system* which revolved around several interconnected factors:

- *just-in-time* supplies and low inventories;
- development of a *skilled and motivated workforce;*
- *quality control* and improvement, including diagnostic discussion
- groups known as 'quality circles'.

4.7.2 Mass customisation

The Toyota story is just one illustration of a quest that can be traced back several decades. The quest, which is still evident today, is to manufacture a range of products on a single manufacturing line. In principle this leads to the possibility of mass customisation, where products can be personalised for each customer, but are still produced in very large numbers overall.

Mass customisation of products involves choosing options from a suite of possible designs, by swapping and changing components and sub-assemblies. If a team is to design for mass customisation then they have to look carefully at the product range which the company offers. Clearly it is necessary to standardise the parts and sub-assemblies which are used. The options are then created by choosing which combinations are included in a specific product.

There is a temptation to think that design of mass market products should be about these huge ranges of products. However, mass customisation comes at a price. Just think of the number of possible variations of a single car model you can purchase. Increasingly cars are specified by customers and then manufactured 'with the customer's name' on them. Similarly, many consumer product manufacturers have followed the same route. Designers respond by using standard components and sub-assemblies which are assembled together like 'Lego' to suit the individual requirements of customers. But is this quite a true picture?

The idea of mass customisation is driven by the two things:

1 the perception that a company can improve its market position by offering a very wide set of options;

2 the technical developments in manufacturing methods and their flexibility. These developments include computer controlled machines, quality control and computerised planning and scheduling of manufacture.

The success of mass customisation depends on the balance between these two, that is:

(a) customer demand for variety from a single product range; and

(b) the investment in flexible manufacturing plant and running it for a large variety of individual products.

This balance is easily disturbed. Suppose the costs of flexible manufacture are large (compared with competitors) then the price of products in the market rises and unless the product is very special, the market share goes down. Suppose, on the other hand, customers put lower value on the wide product range, and are not prepared to pay the market price for the product, then the product range may need to be reduced. However, reducing the product range does not necessarily reduce cost much because the major costs are the investment in machines and in computer systems to control them.

One way out of this impasse is to reduce the investment required for this flexible manufacture. This is increasingly possible for some products, as the costs of making components by rapid prototyping fall. However, for other products this is not feasible at present. What can the designer do to help? The answer is more rigorous standardisation whilst retaining the appearance of variety to the customer. Thus mass car manufacturers have a few 'platforms' (engine sizes, chassis, etc.) on which they base a whole range of cars. They even preserve platforms when new models are introduced.

It is sometimes assumed that the availability of mass customisation has come at little addition to the end price of the product. This only happens when the investment in manufacturing is offset by very large sales. For many industries and products, especially smaller companies, it is not viable to offer mass customisation on a wide scale. For these companies, designing products with fewer options in the product range leads to savings on manufacturing investment and thus offering a lower price on products with fewer options in order to compete with the market leaders.

4.7.3 Lean manufacture

The kind of thinking which looks for appropriate balances is one aspect of *lean manufacture*. I end this section with a quote from a very influential book called *The machine that changed the world* which is based on a five year international study of the car industry.

The lean producer ... combines the advantages of craft and mass production, while avoiding the high cost of the former and the rigidity of the latter. Toward this end, lean producers employ teams of multiskilled workers at all levels of the organization and use highly flexible, increasingly automated machines to produce volumes of products in enormous variety.

Lean production ... is 'lean' because it uses less of everything compared with mass production – half the human effort in the factory, half the manufacturing space, half the investment in tools, half the engineering hours to develop a new product in half the time. Also, it requires keeping far less than half the needed inventory on site, results in many fewer defects, and produces a greater and ever growing variety of products.

Perhaps the most striking difference between mass production and lean production lies in their ultimate objectives. Mass-producers set a limited goal for themselves – 'good enough', which translates into an acceptable number of defects, a maximum acceptable level of inventories, a narrow range of standardized products. To do better, they argue, would cost too much or exceed inherent human capabilities.

Lean producers, on the other hand, set their sights explicitly on perfection: continually declining costs, zero defects, zero inventories, and endless product variety. Of course, no lean producer has ever reached this promised land – and perhaps none ever will, but the endless quest for perfection continues to generate surprising twists.

For one, lean production changes how people work but not always in the way we think. Most people – including so-called blue-collar workers – will find their jobs more challenging as lean production spreads. And they will certainly become more productive. At the same time, they may find their work more stressful, because a key objective of lean production is to push responsibility far down the organizational ladder. Responsibility means freedom to control one's work – a big plus – but it also raises anxiety about making costly mistakes.

Similarly, lean production changes the meaning of professional careers. In the West, we are accustomed to think of careers as a continual progression to ever higher levels of technical know-how and proficiency in an ever narrower area of specialization as well as responsibility for ever larger numbers of subordinates – director of accounting, chief production engineer, and so on.

Lean production calls for learning far more professional skills and applying these creatively in a team setting rather than in a rigid hierarchy. The paradox is that the better you are at teamwork, the less you may know about a specific, narrow speciality that you can take with you to another company or to start a new business. What's more, many employees may find the lack of a steep career ladder with ever more elaborate titles and job descriptions both disappointing and disconcerting.

If employees are to prosper in this environment, companies must offer them a continuing variety of challenges. That way, they will feel they are honing their skills and are valued for the many kinds of expertise they have attained. Without these continual challenges, workers may feel they have reached a dead end at an early point in their career. The result: They hold back their know-how and commitment, and the main advantage of lean production disappears.

(Womack, Jones and Roos, 1990, pp. 13–14)

Exercise 8: Assessing the predictions

Having read the extract above, to what extent do you think the predictions about manufacture have been realised.

What are the changes identified in the area of design?

Discussion

1 Product development times have reduced, but it is sometimes difficult to compare like with like. Investment on tools may have reduced but again it is hard to judge. Quality (in terms of reduced numbers of defects in products) has improved. Inventories in companies are probably lower, but this is sometimes at the expense of other suppliers holding stocks.

2 Quicker product development, higher quality products, greater variety, project team working, altered career paths for designers.

Key points of Section 4

Four types of production system were identified:

- one-off (bespoke);
- batch;
- flow-line; and
- continuous.

Examples of all processes, except one-off, can be found in mass production.

Manufacturing processes can be broadly categorised as joining, shaping and surface treatment. How to create the appropriate shape and suitability of materials are key determinants in process selection. Other factors include cost, cycle time, flexibility and waste.

Alongside other factors, design teams need to think ahead and consider 'design for manufacture' when designing a product. This might include consideration of relevant production system, manufacturing process, and assembly.

Modern manufacturing has adopted and exploited new techniques, approaches and technologies. These include computer-controlled manufacturing, quality control systems, reduced inventories, reduced product development time and new ways of organising the workforce.

'Lean' manufacture characterises the approach of some of the leading international manufacturing companies.

(Learning outcomes 1.3, 1.4, 2.1, 2.2, 2.3, 3.1, 3.3)

5 Computer modelling for design

Tony Hodgson

5.1 Computer-aided design and manufacture

Computer-aided design
The use of computers in a wide range of design activities from initial concept design to tooling for product manufacture.

Computer-aided design (CAD) is now a significant part of modern design practice, and is a key modelling tool that might be used during many different phases of design development. It opens up many opportunities for the designer to work more quickly, accurately and efficiently. In this respect, it is one of the key means by which manufacturers can more accurately meet consumer demands and more quickly launch products onto the market. In order to achieve this, designers need to be aware of the potential of CAD modelling, particularly when it is linked directly with manufacturing or prototyping. It need not be restricted to two-dimensional (2D) drafting or elements of detailed engineering design. Modern CAD software can be used to generate physical detailed three-dimensional (3D) CAD models.

CAD models are most effective when they define a design in great detail, and they can be used for a variety of design tasks including:

- *product visualisation* (through rendered images or virtual reality interactions);
- *product simulation* (through functional analysis or animation);
- *product prototyping* (in many different forms and at different points in the design development); and
- *product manufacture* (by directly linking with strategies to produce the product components or the tooling needed to manufacture these components – referred to collectively as computer-aided manufacture (CAM).

Computer-aided manufacture
The use of computers to control manufacturing processes.

The combined use of computer-aided design and computer-aided manufacture (CAD/CAM) suggests a high level of computer modelling capability and significant investment in equipment, infrastructure and expertise in order to enable its effective use. Nowadays it is necessary for the designer to understand the role of CAD modelling in the design of their product.

This raises four important issues and questions:

1 What CAD modelling (and related) systems are available and how may they be employed to assist design development?

There are a great many different CAD systems that a designer might use to develop and model their product proposals. Typically, these might be solid modelling systems, surface modelling systems (such as solidThinking on the course DVD) or a combination of both (hybrid systems). The selection of a particular CAD modelling system will depend primarily on the level of detail and complexity of the parts to be modelled, and the ways in which the CAD model is to be exploited (through different outputs) by the design team or their client.

2 To what extent can a designer's notion of product shape, form, function (design intent) be captured in a CAD modelling system, and how might this be achieved?

This can be described as the accuracy of the CAD model compared to the designer's intended model. That is, how well can the 'design intent' be captured by the CAD modelling system? There are two key factors to be considered here. First, does the CAD system have appropriate features for the creation and subsequent development of the product/part form and detail? Secondly, does the designer have the CAD modelling capability to be able to use these tools effectively?

It may be that one alternative to the very significant CAD modelling skills required is to 'reverse engineer' complex forms in materials such as rigid cellular foam or clay which can be shaped by hand more easily. A model or existing product can be scanned by laser or touch-probe to create a cloud of points in 3D space, which represent its geometry. These points can then be used to automatically build the surfaces or solids in a 3D CAD modelling system.

3 How can the potential of CAD assist designers with the many detailed design iterations required in various projects?

Experienced CAD users are able to make many changes to their design by revisiting the CAD model and making small amendments very easily, rather than completely redrawing a product or producing a new sketch model/mock-up. This allows the designer to explore new concepts, or variations of concepts, with greater confidence and in shorter timescales, and leads to a more thorough design development. For many designers, the aim would be to have a design that meets required specifications when it goes to market, hence the phrase 'right first time'. CAD modelling allows greater opportunity for designs to be modified, tested, evaluated and reconsidered. In this respect, there is less pressure to get early design iterations 'right', since they can be easily amended at a later phase of the design work.

4 How is the nature of design activity changed by making use of CAD modelling strategies?

It is worth reflecting on how the intensive use of computer-based systems can alter the design approach. The ability of CAD to enable many design iterations in a relatively short period, and to allow designers to easily visualise, simulate and amend designs provides them with a more malleable, even virtual, development system. This has had a major impact on previously perceived design methods.

Moreover, computer modelling may be seen as a means of managing concurrent or simultaneous development of design (and manufacture). In this sense it can be viewed as an enabler of the concurrent design methods and philosophies discussed in the previous section.

5.2 Computer modelling systems

Most CAD systems represent design models on a flat viewing screen and they allow for interaction with the designer in order to define and refine the model. Typically, the system will comprise a series of computer modelling tools that may be used to define lines, forms, volumes and surfaces. Different CAD systems allow different types of model to be generated.

2D wireframe modelling

Two dimensional drawings are used extensively in many manufacturing operations. They are also used to record and store the details of parts. Two dimensional drafted output is generally available from any typical CAD modelling system, but systems in this category limit themselves to the creation and modification of drafted output, requiring designers, and others, to interpret the drawings themselves as 3D parts and products. An example of 2D drafted output is shown in Figure 47.

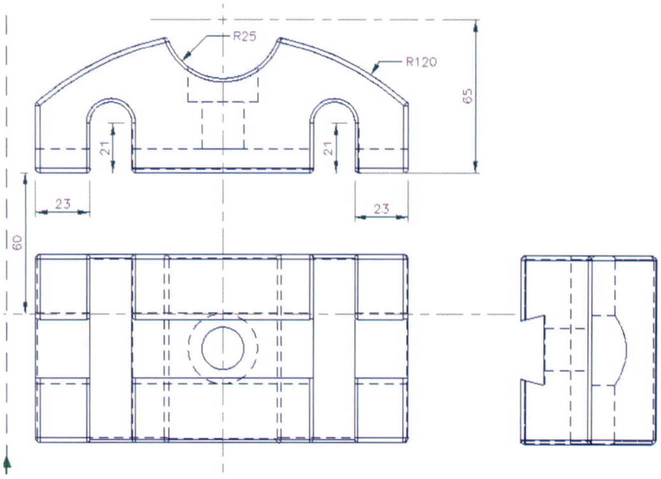

(a) 2D drafted views of a toy car 'body'

(b) The real wooden toy

Figure 47

3D wireframe modelling

Whilst this may appear to represent an object as a three-dimensional form it offers only a limited representation (see Figure 48). For example component edges may not be sufficiently defined for this system to be directly connected with manufacturing systems.

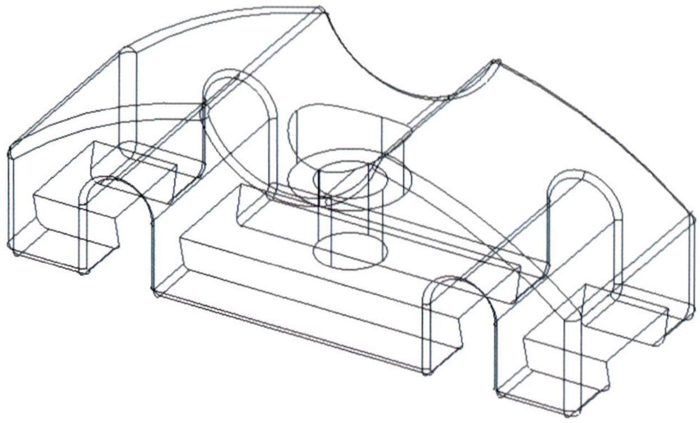

Figure 48 Basic 3D wireframe model of the toy car body

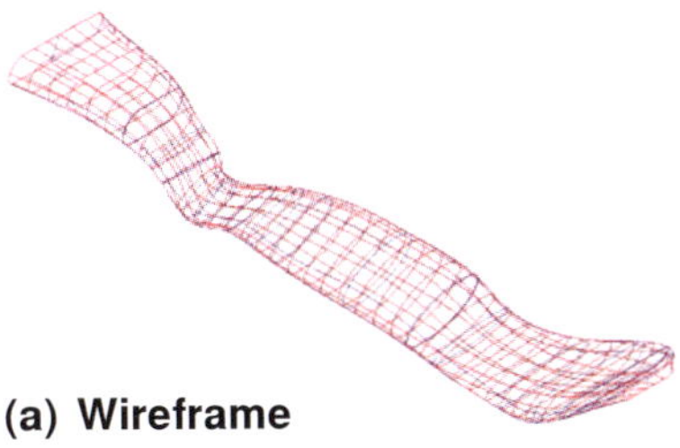

(a) Wireframe

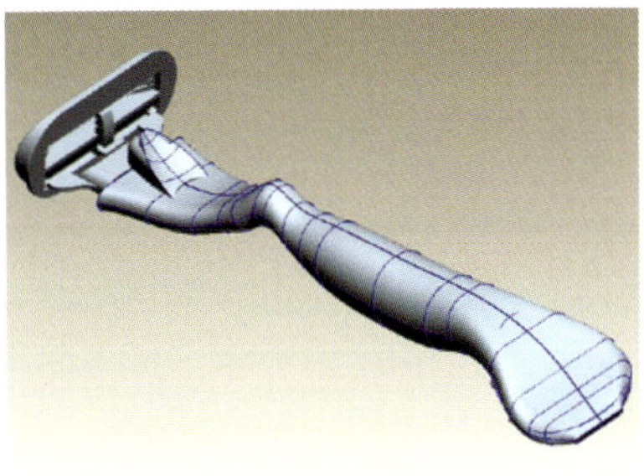

(b) 3D surface model

Figure 49 3D CAD models of a ladies shaver

Some wireframe modelling techniques allow the extraction of 2D drafted output from 3D models defined in wireframe form and some systems allow wireframe models to be given 'surface skins'. These more sophisticated models can provide enough definition of the part geometry to enable some rendered output, manufacture, or part prototyping to be taken directly from the CAD model. For example, 'stereolithography' files (generally required for rapid prototyping systems) may be generated from such systems.

Significantly, these systems allow for modelling of a more complex nature than simple wireframe models are able to define, so 'design intent' may be captured more effectively (see Figure 49).

3D solid and surface modelling

This system is central to a reliable and effective CAD modelling environment. Solid modelling allows parts and products to be created as 'real world' models, attributing material properties in addition to definition of shape and form. This extends the modelling approach so that different parts, components and features can be combined and integrated allowing a detailed 3D model of a complete product to be created. The creation of surfaces, as separate entities to solids, is becoming less commonplace and 'hybrid' surface/solid modellers allow for a seamless combination of both feature types (see Figure 50).

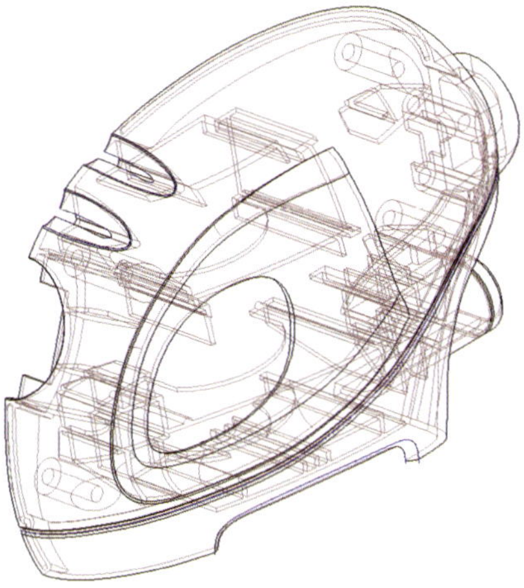

Figure 50 Two images from a combined (hybrid) surface and solid model of an electric sander body shell

Parametric modelling

The defining geometry of features may be linked to other parts of the geometry, or to other features, possibly in other parts. Modifying features in one part can make linked modifications in others. For example, the shape of a button for an electrical product may be linked to the shape of the recess in which it must be located. Modifying the design of the button will automatically modify the recess in the product case. Similarly, it is possible to link the number of spokes in a wheel to its diameter. Larger wheels may require more spokes, and the relationship between them can be easily defined in a parametric modelling system.

Parametric modelling allows the design to be amended quickly and easily, and helps designers to work concurrently on different parts of a product, since each may be linked to other relevant parts in the design.

Feature-based modelling

Features are stored in a library, usually as pre-set parts or feature elements of a product. They are often definitions of material, manufacturing methods and other specifications, in addition to solid or surface modelling data. Such features can help to speed up development of a range of products, or subsequent product models, and can be a particularly effective way of sharing design models between suppliers and their manufacturing partners. The CAD system is especially good for designing new models in a range from standard components.

5.3 CAD in prototyping and manufacture

CAD modelling systems, and their associated visualisation systems, have reduced the need for modelling and prototyping in real materials. Design concepts may be presented by rendering a detailed CAD model in a realistic context (e.g. a toaster presented next to other breakfast items or creating a hand around the shaver in Figure 49). This can provide a very good illustration of how a product will appear. In addition, detailed engineering drawings may be taken from the model, allowing prototyping or manufacturing to be anticipated.

However, despite even the recent advances in stereoscopic and virtual environments, there remains a need to make and test prototypes made from appropriate materials so that tactile, ergonomic and other vital feedback may be evaluated. A new camera, for example, may appear to include all the required elements and specified functions, but until potential clients, users and the designer are able to hold and make use of such a product it will not be possible to fully evaluate the elusive human factors of the design.

In terms of product manufacturing output, some designers expect to have full control over the parts to be made. If the approach is to take engineering drawings and pass them to a third party for interpretation and implementation, then there is scope for error and unwarranted design change. In this context the potential for a manufacturer of parts to work directly from the CAD model provides designers with greater assurance that their design will be implemented as intended.

There are, essentially, two different ways in which prototype output may be derived directly from the CAD model. The first uses cutting, milling, routing and turning operations to shape stock material in line with the part model. This 'subtractive' process uses computer-aided manufacturing systems to automate the more traditional machining operations (as discussed in Section 4). The second builds parts from the 'ground up', in a series of very thin slices or layers. This 'additive' process is typically described as 'rapid prototyping' (RP).

rapid prototyping
Various processes for exploiting digital information, typically from a CAD program, to build up tangible components in a series of very thin layers.

Both approaches require software applications in addition to the CAD modeller, and also some expertise in the use of the appropriate manufacturing equipment. Some of the issues arising from their selection and implementation are discussed below.

5.3.1 Computer numerical control (CNC) machining

CNC equipment is primarily used to manufacture finished parts or to make the mould tools for products, but it also has a role during design development as a means of manufacturing prototype parts and components. CNC equipment removes material from a block or billet – i.e. it is the 'subtractive' process referred to above. Cutting tools are employed, typically in lathes, milling machines or routers. The type of equipment specified can greatly influence the design of a part, since it must be possible to manufacture (or prototype) that part with the chosen technology. For example, complex part shapes can be difficult to produce without the use of expensive equipment. If the intention is to (ultimately) manufacture parts of a product in this way, then such a limitation at the design stage is acceptable. However, if this provides an unreasonable restriction on the prototype development, then designers should consider alternative prototyping strategies, possibly including rapid prototyping (discussed below) which allows more complex and intricate component parts to be made directly from a detailed CAD model.

There are, generally, three different ways in which CNC machining may be used in the manufacture of product parts or prototypes:

1 Production machining

Such parts would have relatively simple shapes and the volume of production would generally be small to medium batch size. For example, parts of 'flat-packed' wooden furniture might be made this way. In this system the CAD model and CAM are usually directly linked.

2 Toolmaking

Tools (for manufacture of large volume parts) are expensive, complex and often unique. For example, the mould tools for a complex polymer vacuum cleaner shell might cost over half a million pounds and consist of several parts.

CNC programs guide the cutting and shaping machines over and inside the form and seek to get it 'right first time' because it's expensive and difficult to correct errors in moulds. Tool design may need to make allowance for any shrinkage that may take place in the mouldings as they cool, cure or set.

3 Design prototyping

CNC machining is also used in one-off manufacturing and prototyping. Examples of this would be the making of block models of a telephone handset or a prototype hose coupling for testing and evaluation.

Prototyping from complex CAD models leads to 'malleable working prototypes'. Thus if designers wish to change the design of a product or component following the evaluation of an earlier prototype then the machining of the subsequent prototype is easily facilitated, particularly if the CAD model and machining operations are linked as described earlier.

5.4 Rapid prototyping

The term rapid prototyping (RP) describes an additive process that results in a physical output from a 3D CAD model. This prototype can be very high in quality so long as the detail shape and form of the components are clearly specified by the designer. Some RP systems lead to fragile models, of use only in visual assessments. Other RP systems lead to robust models of parts, strong enough to be incorporated into working prototypes.

Rapid prototyping is described as an additive process because it builds up layers of material to form the part. The thinner the layers, the more detailed the part can be, but the longer it can take to be made. Most rapid prototyping processes work by using this same basic additive method, they just use different materials and manufacturing technologies to produce the part. The part could be designed using any CAD package, so long as it completely and accurately describes the shape of the product. It is usual to prototype each component of a product separately.

Once a part exists as a CAD file this can be converted into a stereolithography file, (often called an 'stl' after its computer file extension, .stl). The stereolithography file format has come to be a generic standard for all RP and manufacturing processes. Converting the CAD model to an stl file discards a lot of information about the model, (colours, names of components and other information describing the CAD model history), retaining just the geometry and detail of the outside shape. This geometry is described by creating many different sized triangular facets on the surface of the model. Where a model has a flat face, there are only a few large triangles. Where a surface changes a lot, such as at a curved edge, there are numerous small triangles. This is shown in Figure 51.

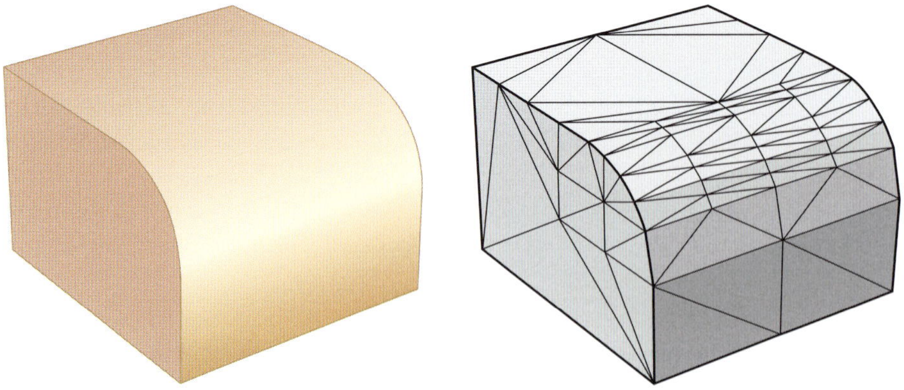

Figure 51 Preview of triangular facets in a stereolithography representation of a simple block

Note the increase in triangular facets where the surface becomes curved.

The stereolithography file is put into a piece of software that slices up the model into very thin layers. The RP model is then physically built up layer by layer – often taking many hours. If parts of the model 'overhang' the base, such that it cannot support itself as one layer is built on top of another, the program will automatically devise a temporary framework to support it.

The final stage is to clean up the model, removing any added support framework. This can leave small spots of material on the surface of the model, which are then cleaned off by more traditional modelling methods and hand finishing.

Use of RP technologies is an attractive means of providing prototype models, particularly if the parts are complex in nature and are already designed in CAD. However, there are a number of issues concerned with the use of RP as part of the design and development of a product, which should be considered.

Costs

Initial equipment costs are generally high, and many companies will choose to have parts prototyped by outside agencies, rather than incur the cost of purchase, maintenance and running of the RP equipment. However, the cost of the RP models compares favourably with the cost of employing model makers to produce the parts, particularly where parts are complex.

Selection of an appropriate output

There are various types of RP process, each utilising different materials and leading to a range of different output types. Each presents its own strengths and weaknesses in terms of model resolution, physical properties, build time and cost. Designers and engineers select an RP technology that best suits the needs of a prototype at a particular time. It is likely that early concepts, which may only need to be evaluated quickly and visually, would be prototyped in a different way to final appearance prototypes which would need to be more robust and include significant product function in addition to representing the final appearance of a design. The examples that conclude this essay illustrate how such judgements may be made.

Rapid manufacturing

Some of the newer RP processes can produce components from the same engineering materials used to produce a final component. This means that a prototype from an RP machine can be used as the final component. At the moment it is not feasible to use an RP machine to produce components in great numbers, as it takes too long to make each component. But future rapid manufacturing systems will be able to produce components much more quickly and will be fully automated, with robots to move the completed component between different manufacturing equipment and to ensure that the machine has enough material to make more components.

Such integrated manufacturing forms part of Flexible Manufacturing Systems (FMS), which unlike current manufacturing processes, will be able to make completely different components from hour to hour. This would provide potential for parts to be customised to suit the needs of specialised groups, or even individual users. This opens up further opportunities for designers who can design components and parts that are not constrained by more traditional manufacturing processes and who can react to client and user needs much more quickly than before. RP technologies can, potentially, provide a significant contribution to the goal of 'low cost' mass customisation discussed in the previous section.

Table 3 provides a comparison of five rapid prototyping techniques. The three most significant techniques are discussed in more detail in the following section.

Table 3 An overview of rapid prototyping techniques

General features	Process				
	Stereolithography addition (SLA)	Fused deposition modelling (FDM)	Laminated object manufacture (LOM)	Selective laser sintering (SLS)	Solid Object Printing (SOP) – Wax Modelling
Typical maximum part size (mm)	508 × 508 × 609	609 × 508 × 609	812 × 558 × 508	380 × 330 × 457	254 × 203 × 203
Speed	Average	Good	Good	Average / good	Good
Accuracy	Very good	Fair	Fair	Good	Good / very good
Surface finish	Very good	Fair	Fair / poor	Fair	Good
Strengths	Simple process to implement. The ability to manufacture parts which would be impossible to conventionally produce in a single process. Virtually any geometrical shape can be produced. High resolution.	Quick process. Cheap generation of 3D models. Easy to use. No toxic chemicals or lasers. Little waste material. Materials can be changed quickly and easily. Models require little finishing when used as functional models.	A variety of organic and inorganic materials can be used: paper, plastic, etc. The costs are relatively low. Relatively slow manufacture time.	Various materials can be used: polycarbonate, PVC, nylon, croning sand, alloy, investment casting wax. No post curing is required. No support structure is required.	Fast manufacture time. Inexpensive (with respect to other RP techniques). Good for investment casting.
Weaknesses	Requires post curing. Support structures are often required. Toxic resin with high health and safety risks.	Restricted accuracy for detailed work due to extrusion head diameter.	Stability of models. Only suited to simple models. Hollow parts cannot be easily manufactured. Models need to be sealed.	Possibility of toxic gases (especially with PVC). Surface finish may feature stair steps.	Support removal is tricky. Relatively fragile parts.
Typical system cost	Very high	Low	Meduim	Very high	Medium

Note: These figures provide only a general guideline; there are many different RP machines available that could offer different results

5.4.1 Stereolithography addition (SLA)

Stereolithography is probably the best-known rapid prototyping method. The technique is based on a 3D printing process that uses a vat of photosensitive resin and a vertically moving elevator platform. A laser beam, directed onto the surface of the photosensitive resin via mirrors, 'prints' the pattern of the current model layer by hardening the photosensitive resin (Figure 52). The platform then moves down by a layer thickness so the next layer can be 'printed'.

The process is relatively quick and accurate and can produce relatively strong components. The initial cost of the equipment, and cost of the parts it produces, is very high.

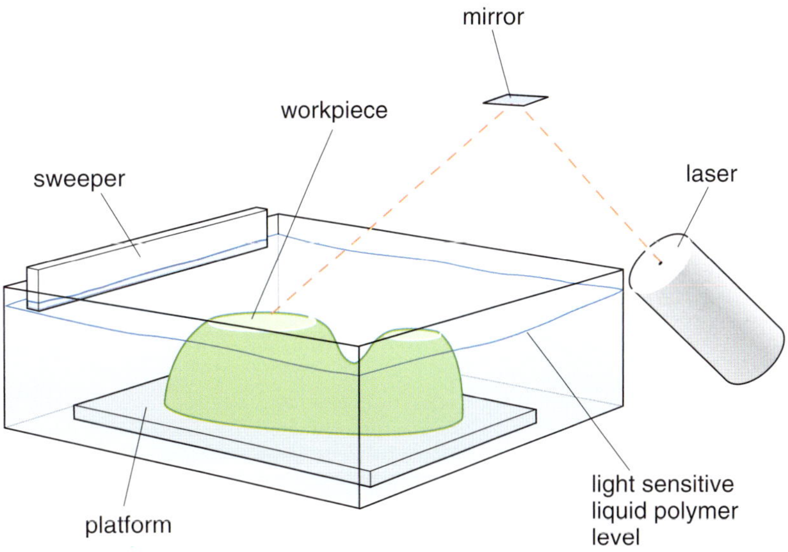

Figure 52 The SLA method of producing prototypes

Figure 53 (overleaf) illustrates the process in more detail. The part-built model appears just below the level of liquid polymer resin. The laser then 'prints' the shape of the next slice of the model onto the surface of the photosensitive resin (step 1). When the photosensitive fluid is exposed to the UV laser beam the area affected solidifies by a low energy absorption process (step 2).

When the laser has finished 'printing' the layer, the elevator platform moves down by the thickness of that layer (step 3). A sweeper moves the liquid resin across the surface to cover any exposed areas (step 4). The process then begins again and continues until the whole model is complete.

The fragile resin model that results is removed and is cured to fully harden it.

Because the model is built up in a series of layers, its surface contains a series of steps. To achieve a smooth model, fine glass paper can be used to remove these 'steps' from the model.

Models can be primed and painted to provide an appearance model. Some SLA resins once cured are strong enough to be used as parts in working prototypes.

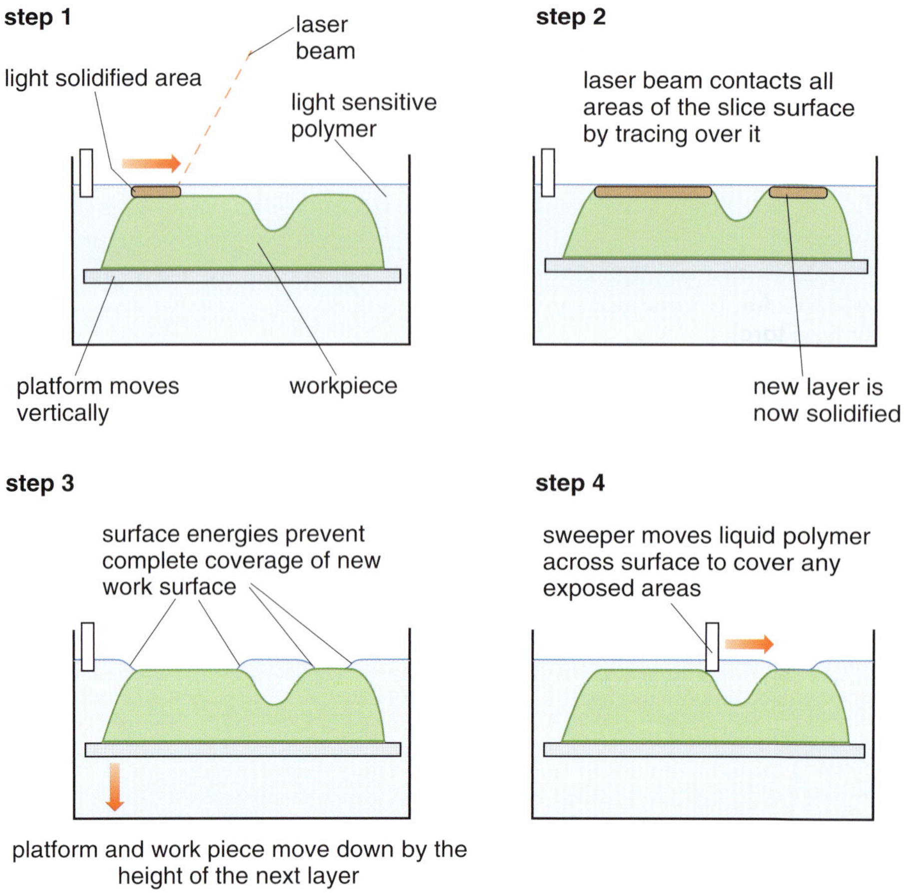

Figure 53 The RP build process

Because of the liquid environment in which the models are produced it is often necessary to incorporate support structures to the model (see for example Figure 54). These support structures are used to hold parts of the model in place and to maintain the structural integrity of the model while it is being manufactured.

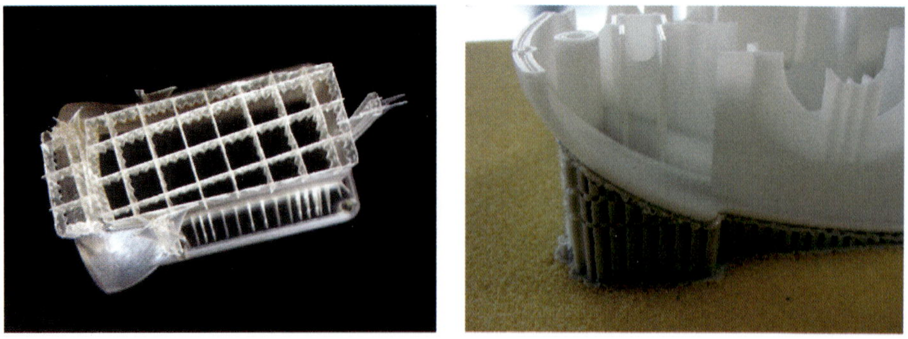

(a) SLA model

(b) FDM model

Figure 54 Two RP models showing the contrast in output of two methods

Note the support structures on both models.

The orientation of the model during building is an important factor when considering surface finish, build time and cost. If the model is orientated with its longest axis vertically then it will take longer to build but will have fewer 'steps' and thus a smoother surface finish. However, this orientation costs more to produce due to the height of

Figure 55 An SLA model of a prototype torch top shell

the model and the number of layers required. If the model is orientated with its longest axis horizontally, the build time will be less (due to the lower height the machine is required to reach) but the surface finish can suffer due to the number of 'steps' in the model. Some models require considerable manual finishing.

This is one of the most expensive rapid prototyping processes available at present. Industry standard machines cost (in 2004) from £350,000 and upwards, with a vat of resin costing from £20,000. Table 4 summarises the relative advantages and disadvantage of SLA rapid prototyping.

Table 4 An overview of SLA

Advantages	Disadvantages
Simple process to implement	Requires post curing
The ability to manufacture parts which would be impossible to conventionally produce in a single process	Long term curing can lead to warping
Virtually any geometrical shape can be produced	Parts are relatively brittle and have a tacky surface finish
Highly accurate physical reproduction of CAD model	Support structures are often required
	Requires a detailed sequence of processes
	Post processing labour needed to clean up the finished component
	Toxic resin with high health and safety risks
	Running costs are very high

5.4.2 Fused deposition modelling (FDM)

An example of an FDM model is given in Figure 54(b). An FDM machine is basically an arm that holds one or two small heads for extruding a polymer. The extrusion head moves back and forth along a platform building up a 3D model by feeding heated polymer wire through the extrusion head (see Figure 56).

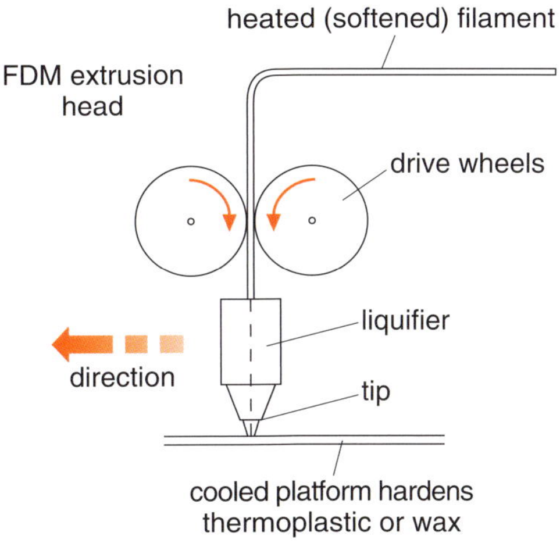

Figure 56 The FDM process

The FDM machine works in the same way as a CNC machine by moving the extrusion head along the axes of the FDM machine. Molten material is extruded from the head and hardens immediately in ultra thin layers to form a complete 3D model.

Some FDM machines have two extrusion heads. One builds up the model using the molten material while the other extrusion head uses a different material that produces the supports required for complicated models. These supports can easily be removed after the model has been completed.

There are very few problems with FDM manufacture. Little finishing is required to FDM models except where support structures need to be removed. Finishing around these areas can be carried out by hand with fine glass paper.

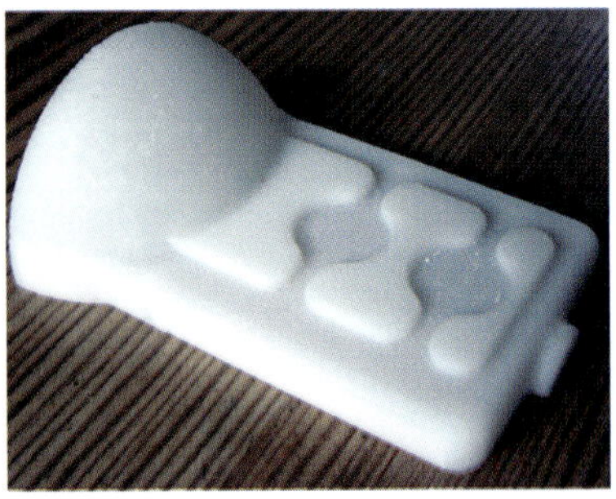

Figure 57 An FDM model of the torch top shell

In relation to the SLA and selective laser sintering (SLS) machines, FDM proves to be very cost effective. Typical industrial machines cost around £50,000 although smaller office-based machines are available for around £25,000 thus enabling design offices to exploit rapid prototyping alongside more traditional modelling techniques. Figure 57 shows a typical FDM model. Table 5 summarises the advantages and disadvantages of the process.

Table 5 An overview of FDM

Advantages	Disadvantages
Quick process	Restricted accuracy for detailed work due to extrusion head diameter
Cheap generation of 3D models	
Easy to use	
No toxic chemicals or lasers	
No waste material	
Materials can be changed quickly and easily	
Models require little finishing	
Accurate, functional models	

5.4.3 Laminated object manufacturing (LOM)

The LOM machine takes data from the CAD model and cuts out each layer from a roll of sheet material, using a laser. These sliced layers are glued together to form the model, which is built on a movable platform below the machine (Figure 58). Imagine cutting out many hundreds of paper shapes – each one representing a particular section of your chosen product, then gluing them all together to get a 3D version of your product. Well, that's how LOM works.

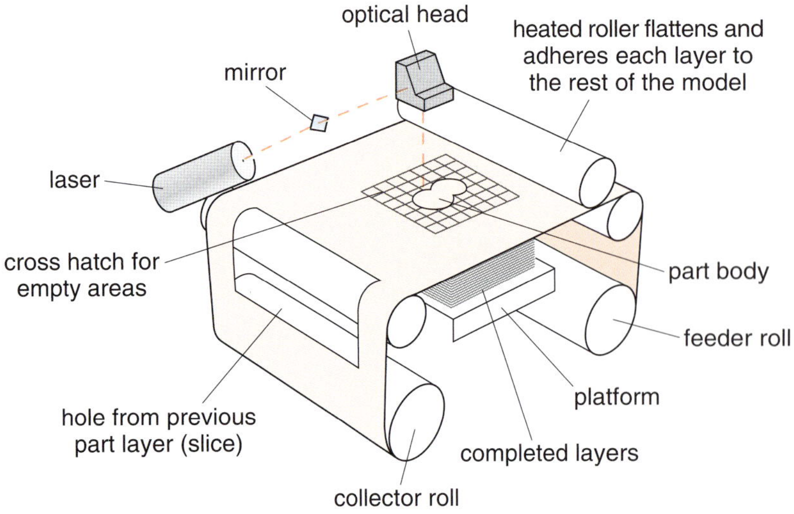

(a) The LOM process

(b) Demonstration model of a LOM part built from many layers of paper

Figure 58

The control computer that runs the LOM machine is capable of slicing a 3D CAD model into thin 2D cross sections. The thickness of each cross-section is set to be the same as the thickness of the material being used.

The material used is typically a glue-backed paper, which is stored in sheet form on a roller on one side of the machine. The material is fed from this roller and passes over the central platform. On the other side of the machine is another roller that collects the material once it has been used.

A laser is used to carefully cut out the appropriate layer of the model. A heated roller then moves across the machine adhering the sliced layer onto the 3D model that is being built below. This 3D model is formed on a movable platform that, once a new layer has been added, moves down ready for the next layer. Once the platform has moved down, the material moves along, via the rollers, ready for the next slice of the model to be cut.

The laser cuts up or 'hatches' unwanted areas of the LOM material and this makes it easier to remove once the model has been made. Surface finishing can easily be achieved by using some glass paper to remove any 'steps' in the surface of the model (see Figure 58 (b)).

The main problem to this method is that the LOM process can only make solid parts (i.e. hollow parts, like bottles, cannot be built via the LOM process because the paper inside the model cannot be removed). The process also struggles when producing vertical thin walls on models (see Table 6).

Table 6 An overview of LOM

Advantages	Disadvantages
A variety of organic and inorganic materials can be used e.g. paper, plastic	Stability of models depends on the bonding strength of the glued layers
The costs are relatively low	Only suited to simple models
Relatively rapid manufacture time	Hollow parts cannot be manufactured
	Models need to be sealed against moisture

5.5 Studies of the use of rapid prototyping

5.5.1 A food preparation product

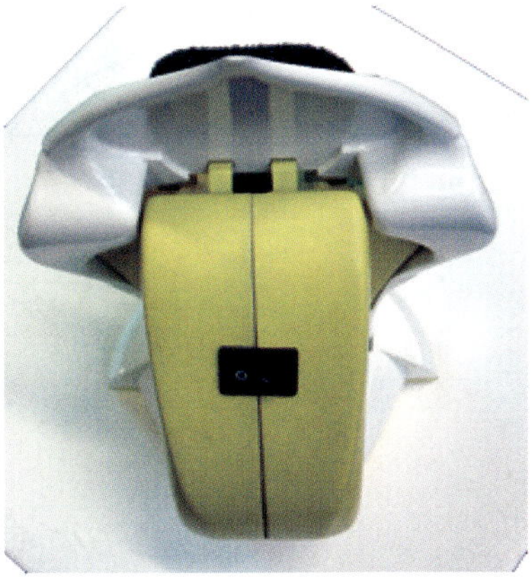

Figure 59 Working prototype of the food preparation product

This example illustrates the use of a wide range of RP output at different phases of design work. As well as considering its function for the preparation of vegetables the designer of this product was aiming to associate it with creativity and fun in the kitchen. The design also exploits a new textured polymer. The working prototype can be seen in Figure 59.

A textured pad can be interchanged (together with some other parts) so that the product peels, cuts or shreds vegetables. The yellow coloured drive unit is battery powered and snaps into the white part, which is used to collect and hold water to keep the vegetables clean. The textured pad can just be seen as a black part towards the top of the unit.

The design started by analysing the potential market for such food preparation products and considering the needs of users, helping the designer to generate initial ideas, concepts and design specifications.

During this time, some mechanical parts were modelled in CAD, and evaluated through mechanical simulation, but it was still felt that there was a need to physically model the mechanism in a wood and aluminium mock-up. The shape and form of the different parts were also developed using a combination of sketch work and CAD modelling, and once again it was felt necessary to check some parts by rapid prototyping them in wax (see Figure 60). This helped the designer to consider the appearance of their design, the comfort, and related ergonomic factors of the hand-held parts in particular.

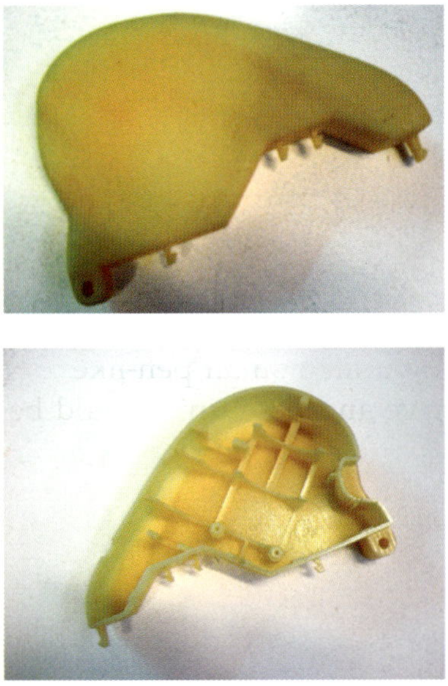

Figure 60 Top and bottom views of the wax RP model

These parts could have been modelled in foam, either by CNC or by hand, but the wax RP model offered an accurately formed prototype part that very closely followed the CAD model and the designer's intended design. It required minimal finishing and was quickly achieved using the stereolithography file generated from 3D CAD modelling software. The ability of the system to build internal features of the casing was particularly beneficial. Although the model was quite fragile it allowed for quick and easy checks for fit and location of mechanical parts. Other types of model were used to evaluate product function.

A number of wax-modelled prototypes were produced during the concept development phase of the design development, alongside continuing sketch and CAD modelling work. The use of RP output that is easily and quickly achieved enabled the design to be developed more quickly and effectively through a number of early design iterations.

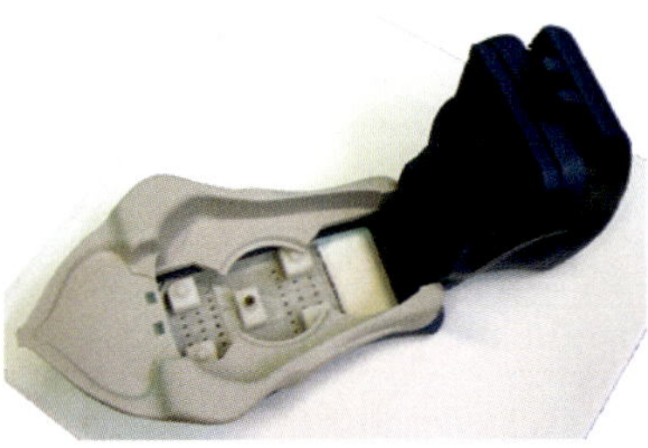

Figure 61 Finished FDM parts

As the design progressed, it became necessary to treat components more roughly and to fit parts together as more detailed design development was undertaken. The designer used FDM for base parts of the product, so that she could evaluate how well the motorised unit would locate in the base (Figure 61). She chose to finish these parts by hand in order to clearly communicate her design proposal to her client.

The hand-held part comprises of two units. The lower part contains a mechanism for slicing, grating or peeling, and the top contains a small motor and batteries. The top part hinges upwards, and can then be removed to aid cleaning. Whilst the designer developed these parts using the FDM output to prove her design form and function, the quality of surface finish and detail was not considered to be good enough for an appearance prototype (a prototype that looks *and* operates as the final product). Following some minor redesign after evaluating the FDM models, she used SLA resin output as the basis of a master model (Figure 62) for subsequent vacuum casting of parts in a high quality resin.

Figure 62 Checking the fit of the SLA 'master model' before vacuum-casting

The SLA models required less finishing than FDM models, and were vacuum cast with appropriate colouring pigment so that painting was not required. This resulted in an appearance prototype that was robust enough to be used by a number of people in a typical food preparation context. The resulting evaluation provided very effective feedback on the acceptance and operation of the product.

In the next example, greater distinction is drawn between two different rapid prototyping processes. The design is, essentially, a redesign of a glue gun that was to be aimed at the 'hobbyist' and 'craftsperson' who wanted more sophisticated control of the glue being dispensed. The final design integrates mechanical, electronic and visual elements in order to meet the intended design specification.

5.5.2 Glue gun

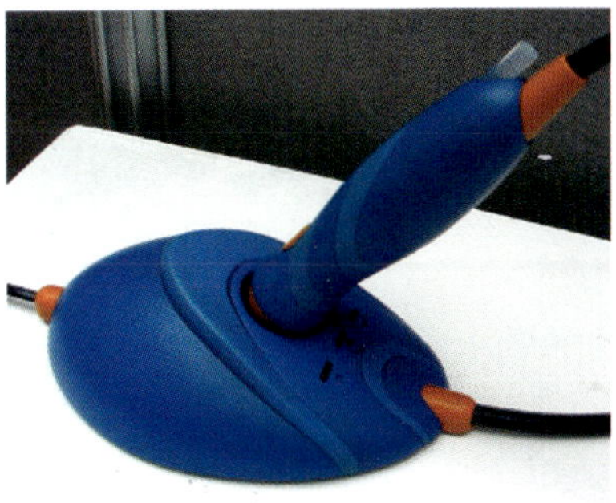

Figure 63 Completed appearance model, in SLA resin, of the prototype glue-gun and base

Early design development highlighted significant problems in creating a motorised glue delivery system within a small pen-like product. It was decided that a separate drive and base unit should be used (Figure 63). As well as the various models made throughout the design process, two different prototypes were made towards the end. Firstly, an appearance prototype that represented the appearance and 'feel' of the final design, and, secondly, a functional prototype that represented the functional operation of the product – a key part of the 'added value' intended for this new product.

The designer developed two detailed prototypes (see Figure 64), in parallel, using CAD modelling to represent each as a detailed assembly of parts. One prototype was to represent the new glue gun as it would look, requiring custom designed electrical and mechanical parts (the appearance prototype). The other prototype included a similar (though slightly different) 'gun', but a much simpler 'drive unit' that would house a prototype motor drive (from a cordless screwdriver) and prototype electronics circuit (for the functional prototype). The designer's intention was to deliver the glue with varying speed, to sense when the user had finished and retract the glue stick to prevent it dripping, and to vary the power to the heating element depending on the intended glueing operation. The product required significant development and so a larger drive unit, with substantial space for further work was built using the SLS system. Figure 64, shows the two different prototypes side by side, and their differences are quite apparent.

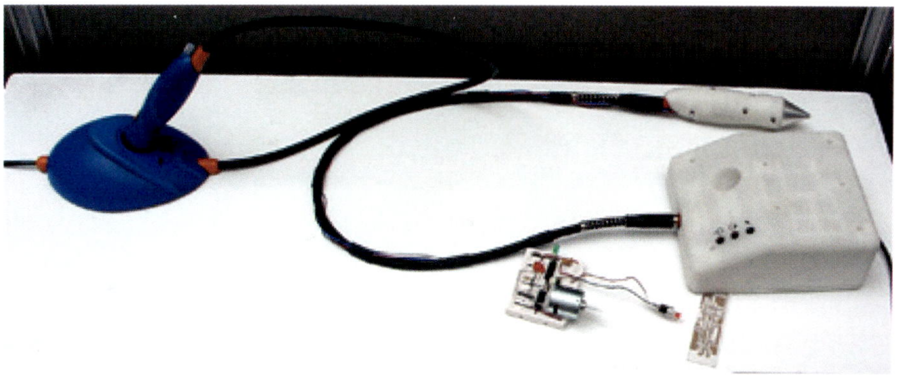

Figure 64 The SLS functional prototype (right) next to the appearance prototype

The appearance model of the final design was prototyped from SLA parts that were finished and painted. Although not as strong as the SLS model, the purpose of this prototype was only to illustrate form and appearance, and complemented the functional prototype well. No attempt was made to finish the surface of the SLS parts used to make the functional prototype.

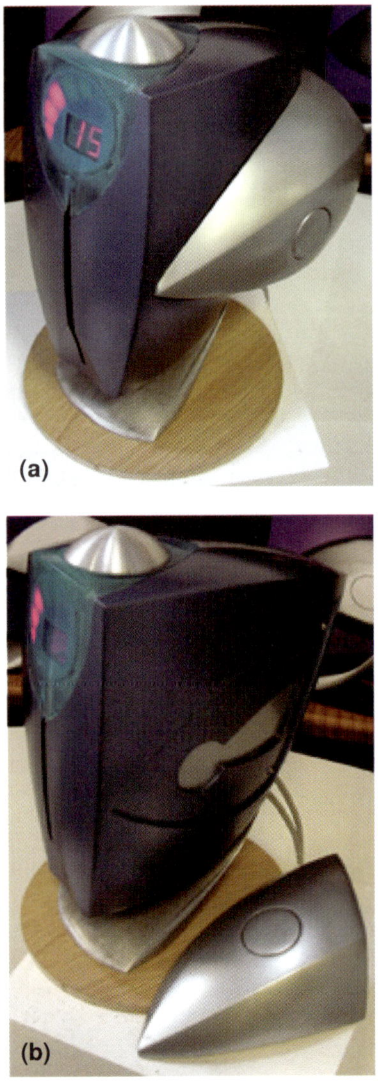

Figure 65 Images of the completed working prototype, with one speaker pod attached (a) and with it removed (b)

5.5.3 CD player

Like the product for food preparation, this prototype CD player used a variety of different RP processes at different phases of the design development work, all linked to significant CAD modelling capability with 3D CAD modelling software. The brief for this product stipulated a domestic CD player where the speakers could be located remotely, and which could be used to control the CD player volume and track.

This CD player consists of a base station that includes a battery charger for the speakers. There are two speaker 'pods' that include audio receivers and control transmitters, together with batteries and NXT sound exciters so that the speaker case acts as a sound board. This allows the units to be used in wet environments, like showers, since the speaker 'pods' can be completely sealed (see Figure 65).

The designer made use of wax models whilst proving the fit of 'pods' to 'base', and used a number of different design iterations to prove the initial concepts for shape and form. The product functionality took longer than anticipated to complete, so instead of using SLA models and vacuum casting to produce prototype polymer shells, he chose to use FDM output and then to finish and paint the surface instead. He considered that the extra time needed to finish the FDM surface would be less than that required to make the necessary silicone moulds and cast the resin parts.

Interestingly, the base unit contained a cast aluminium part as a central 'spine' about which two plastic mouldings were fastened. Although he considered modelling this by hand (which would have sufficed for such an early prototype) he used a wax model of the part to have an investment casting, in aluminium alloy, made as a one-off part (see Figure 66).

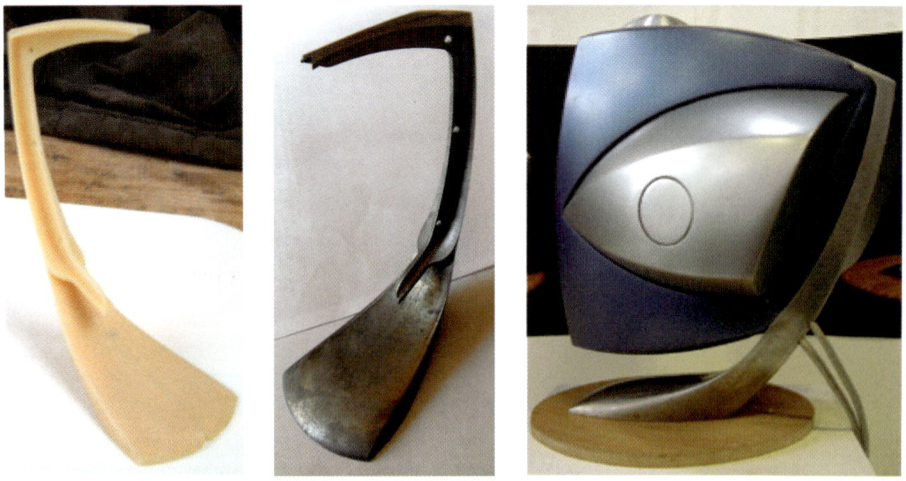

Figure 66 Wax model and aluminium casting of the CD player 'spine'.

This provided a rigid framework onto which the less substantial RP parts could be fastened, and so the CD player could be effectively evaluated as a working, functional prototype.

5.6 End note

In this essay Tony Hodgson has explored the potential and the actual use of computer models in modern product design and development. He has revealed the considerable power in computer-based modelling and technology. But far from seeing the demise of three-dimensional constructions in the design process today, he has shown that systems for supporting computer-aided design actually enable designers to create a wide range of three-dimensional outputs which enable them to evaluate designs at various stages in the process. Rapid prototyping would appear to be an invaluable aid to design practice today and the case studies of design and designing that you will read in Block 6 will explore this further.

Digital models are valuable in many ways. They can be easily stored and modified; they can be distributed to other members in a design team via file sharing systems or simply sent as attachments to e-mail. They can be used to explore strengths and weaknesses in components (e.g. via finite element analysis) or they can be run over a period of time so that the consequences of design decisions can be seen before any great time or money is invested in making a physical model. Sometimes designs are just too big or complex to prototype. Products such as new buildings or passenger aircraft will be developed and proven largely as computer-based models supported by selective building and testing of key components.

The field of product design differs from these large projects in a number of ways. The look and feel of consumer products are vital aspects of their sucess. Very often, the only way to get early feedback on these aesthetic qualities from potential users is to provide them with models, visual prototypes or fully working representations. Product design, more than most other design professions, requires the fast construction of 3D models. The ability of modern rapid prototyping systems to allow designers to create tangible forms from digital models means that we still see the making of both 2D and 3D models. Also, note that many of the components that are built on the various PR systems still require a great deal of hand finishing and painting if they are to visually represent a manufactured part. One of the major breakthroughs has been the ability for RP systems to produce components sufficiently quickly, accurately and with high strength for them to be used in working prototypes. As Tony Hodgson points out in Section 5.4, this offers huge opportunities for future flexible manufacturing systems. We see the use of RP technologies in fashion design, packaging design, furniture design and many other specialist design fields. But it is probably in consumer product design that RP technologies have had the most impact. Most product design work is dealing with relatively small components. These component sizes suit the limitations of the various types of rapid prototyping machines discussed in the essay. This together with the need to see, feel and examine tangible models means that we are likely to see the widest exploitation of RP tools in the product design industry.

Now would be a good time to complete Lesson 3 of solidThinking in the *Block 5 Supplement* if you haven't already done so.

Key points of Section 5

Computer-aided design (CAD) is a term embracing a wide range of digital tools and processes. CAD can be found in the design processes of most modern mass-produced and batch-produced products.

CAD offers designers and design teams a convenient way to model, test, evaluate and communicate ideas. CAD models can be tested safely and accurately. CAD models can be cheaper to produce when compared to a working prototype.

Increasingly, CAD is used to reduce the time it takes to convert a design idea to a manufactured product. Rapid prototyping (RP) describes a variety of techniques that can create three dimensional models from digital models held in the computer.

Some rapid prototyping systems produce fragile models in paper or wax and these might only have a temporary role in design decision making. Other RP systems produce robust components suitable for integration into working prototypes of products and thus they enable user trials or long-term testing.

Far from seeing the end of physical modelling in design, RP technologies have demonstrated the important place of 3D modelling in modern design practice – particularly in fields such as consumer product design. CAD today reveals a close integration of 2D and 3D outputs.

(Learning outcomes: 1.4, 2.4.)

6 Sustainable design

In previous blocks we identified sustainability as a critical function for an increasing number of products. Designing for sustainability is difficult because a product may have many social, environmental and economic consequences or impacts during its life cycle and these consequences are difficult to predict accurately whilst the product is being designed.

Previously we have divided a product's material life cycle into five main phases:

1 raw material extraction;

2 manufacture;

3 distribution;

4 use; and

5 end of life disposal, recycling or reuse.

During each phase the product has social, environmental and economic impacts. One of the challenges for design is to predict the impacts of a product at each stage in its life. Clearly this is more achievable for some types of products than others but, generally, assessment becomes more accurate as more of the details of the product are determined. For example, environmental impacts will depend critically on the materials used in the product as well as on how the product is manufactured. However, even when the design is finished and the product is on the market some of the impacts may still be uncertain, particularly the social and economic ones.

One way to think about impacts is to consider the people who are affected by a product. Although a product will be designed to meet the needs of potential users it will also have impacts throughout its life cycle on a wide variety of other people. These people, who are not direct users of the product also have needs which may be compromised by the product. For example, waste products of manufacture may be harmful to the health of people in the region of manufacture or the over-use of scarce natural resources may mean that future generations will suffer economically. For this section we will think of needs as covering both the direct needs of users and the more indirect needs of people affected by the product during its life cycle.

A design is sustainable if it meets the wide spectrum of needs – social, environmental and economic. But if we are going to design products which are sustainable we need to be able to assess how well a design might meet this spectrum of needs.

An appropriate concept is essential for a sustainable design. For example deciding at concept stage to construct a new chair using a biologically based composite obtained from cropped plants is bound to change impacts in raw material extraction and disposal phases of the life cycle when compared to more conventional chair designs.

As we move to detail design we can start to make some more accurate assessments of sustainability. Decisions made at this stage are critically important in creating sustainable designs. However, sustainable design is only one aspect, although a critically important one, of the more general issue of sustainable development.

6.1 Sustainable development and design

sustainable development
Development that meets the needs of the present generation without compromising the needs of future generations

The World Commission on Environment and Development in a report *Our Common Future* (Brundtland, 1987) defines sustainable development as 'development that meets the needs of the present generation without compromising the needs of future generations'. Designed products make a large contribution to development, from power generation equipment, to cars and food. And as we have seen products have considerable impacts on a wide variety of people throughout their life cycles.

Extraction of raw materials can cause specific pollution and other harmful by-products for those living close to the point of extraction, as well as more general impacts on the quality of life such as visual amenity and increased traffic. Manufacture has impacts on the economics of people and regions, the working conditions of employees, safety and environmental quality for local populations. Harmful by-products of manufacture have been strictly controlled in many countries in order to meet the needs of those living and working nearby. However, it cannot be assumed that even these controls are in force wherever manufacture takes place. Designers and companies have some obligation to limit the use of potentially harmful substances and for monitoring exactly how the products are manufactured and raw materials sourced.

Wider environmental impacts across the world relate to pollution, carbon emissions and potentially harmful wastes. Distribution of products affects those living near roads and airports, for example. Safe disposal of waste and recycling processes will have impacts on local or national populations.

A product may have impacts long after its life. Waste and pollution may have cumulative effects which lead to the needs of future generations being compromised. Convenient supplies of raw materials might have been exhausted or perhaps potential sources of food reduced by contamination.

It should be emphasised that although it may not be too difficult to list a multitude of needs associated with any particular product it is extremely difficult to take them all into account when designing. Given the number and variety of potential needs, and the uncertainty of impacts of the product, balancing those different needs and impacts, both now and in the future, is a daunting and challenging task for designers.

Two approaches can help. First, modelling the needs and second developing special tools to analyse the impacts of a product. The first approach seeks to identify the key needs to be addressed in particular products. Like many other forms of modelling this is a process of simplification. Regulation and legislation can help do this. For example progressive emission regulations for diesel engines provide the main driver for new designs in this industry. We outlined a short case study of an engine manufacturer in Section 5 of Block 4. The second approach aims to help designers understand the impacts of their decisions better. We will discuss an example of a tool for assessing impacts later in this section.

I make another point to illustrate the complexity of the issue of sustainable design. The sustainability of one product may be dependent on other products. For example when waste from manufacturing one product is used as raw material for another. Further, sustainable design is not a product-by-product 'fix' but rather it concerns establishing a long term balance as part of a manufacturing strategy.

Some companies and design consultancies take a systematic approach across the different products they design. Rather than approaching each design task separately, they adopt common policies and strategies across the company. This serves two purposes. First it allows lessons for sustainable design to be transferred between products and second it creates a way of working which focuses attention on the widespread and long-term needs for sustainable design as well as the narrower and immediate needs of the product user. To illustrate these policies and strategies for sustainable design I will examine briefly those adopted by one company you are familiar with in this course – Philips.

6.2 Sustainability and design: Philips

The approach adopted by Philips to sustainable design is summarised in their *Sustainability Report 2002*, published in February 2003. This is a corporate report not a third party independent document. However, it is an audited report so its facts, figures and claims can be verified independently.

The *Sustainability Report* gives an insight into the importance that Philips attaches to ensuring that its activities are sustainable. It indicates what Philips is doing to meet this goal. The whole report is some sixty pages long and I am going to present some quotes which should give you an idea of the activities and future intentions of the company at the time the report was written.

On the course DVD there is a video sequence related to this section entitled 'Philips: design development and sustainability'. Included among the 'extras' to this sequence is a copy of the *Philips Sustainability Report 2002*.

Philips takes a positive and active approach to sustainable development. It does not see sustainability as a constraint:

> We view sustainable development as an opportunity for innovation, enhanced brand reputation and wealth generation, while enriching people's lives and continuing to earn our license to operate.
>
> (Philips, 2003, p. 7)

The main body of the *Sustainability Report* is presented under three main areas of 'responsibility' that Philips perceives itself having. These are

1 economic responsibility;

2 environmental responsibility; and

3 social responsibility.

These correspond to the three elements of sustainability introduced earlier.

The economic responsibility centres around achieving sustainable economic growth of the company:

> Pursuing sustainable development is less an option and more a necessary part of doing business. Many people are turning to the concept of sustainability to identify well-managed and future-oriented companies to invest in.
>
> (p. 19)

The report also identifies the key area of environmental responsibility. A specific programme called 'EcoVision' concentrated on controlling and reducing environmental impacts of products over their material life cycles. In particular the report identifies reducing impacts in the *manufacturing* and *use* phases of product life cycles. The following indicates the general direction of this approach in reducing impacts in manufacture:

> Under EcoVision 1998–2001 the company's accomplishments on the marketing side included integrating Eco Design into the product creation process with a drive toward developing Green Flagship products and a 21% reduction in packaging materials, compared with the reference year 1994. In terms of manufacturing, achievements included reducing actual waste by 60% and water by 50% as well as improving energy efficiency by 24%, also compared with 1994.
>
> We raised the bar in 2002 with the ambitious goals of our latest four-year environmental action program, EcoVision 2002–2005. Progress under this program will be measured against the reference year 2001.
>
> (p. 27)

As regards impacts in use the report gives examples of products and compares them with competitors' products:

> Compared to the average of its three closest commercial competitors the Philips 32PW8717 (a TV with 32 inch wide screen and 100 Hz tube) weighs 8% less, is 33% more energy efficient and uses 15% less packaging.
>
> (p. 31)

Remember this is not a technical document. It does not set out to explain and define technical terms. It describes the general directions of the policies and strategies that Philips has adopted for sustainable development. There are many questions on matters of detail you might ask about the statement above on the Philips TV such as: 'what is the energy efficiency of a TV?' or 'what is a commercial competitor? – a competitor in terms of price or performance or an aggregation of both?'.

In the report facts and figures are given both in comparative terms – percentage improvements in product performance and environmental impacts – and absolute terms, e.g. for CO_2 emissions, waste and water consumption. It is interesting that in this section of the report particular attention is drawn to Philips' requirement for over 17 million cubic metres of water in 2002. This is mainly 'ultra pure' water for semiconductor manufacture. The product designer will have little impact on this, although manufacturing engineers can redesign manufacturing processes to conserve and recycle water.

Finally, the *Sustainability Report* considers social responsibilities including health, safety and education of the workforce. Obviously changes in company organisation have social implications:

> Globalisation, new economies of scale and changes in regional labor costs and resources have contributed to Philips presence in Europe shifting away from manufacturing toward R and D (research and development), marketing and sales and knowledge based enterprise ... On the other hand, manufacturing is concentrated more and more in Eastern Europe and Asia, both in Philips' own facilities and with subcontractors.
>
> (p. 45)

From these extracts from the *Sustainability Report* it is apparent that the extensive scope of activities across the company to deliver sustainable products requires considerable coordination between different parts of the company – particularly design and manufacture. The widespread and complex needs to be satisfied by a sustainable design appear to require an equally complex response from the whole company.

Sustainable design is a major challenge. There seem no easy responses to complex needs now or in the future. The Philips *Sustainability Report* indicates the type of response that can be made. However, I would like to emphasise two points. First, as a corporate report this publication looks to the success and survival of the company. Sustainable design is clearly central to this aim. Philips are concerned, both with specific activities as well as company organisation, to respond to requirements for sustainable design. Second, the wide and complex needs for sustainable design are difficult to satisfy. There are no easy solutions and the initiatives of pioneering companies like Philips are just the first steps. However, although the complex needs for sustainable design and their long timescales present considerable challenges this is not really something new for designers. They have always dealt with trying to balance complex and conflicting needs, not always clearly defined, and with long term consequences.

SAQ 6

Outline briefly the three main areas of responsibility that the Philips *Sustainability Report* describes.

In order to identify more clearly the complexity, breadth and timescale of the needs to be met by sustainable design I will now look at an entirely different type of report which explicitly concentrates on environmental issues in product design.

6.3 Environmental issues in product design

Industry and Environment is a quarterly review from the United Nations Environment Programme (UNEP), Division of Technology, Industry and Economics. *Industry and Environment* contains articles by independent academics and writers. Note that 'The contents of ... [the] review do not necessarily reflect the views or policies of UNEP, nor are they an official record.' The article I will review was written by Bert Bras, at the School of Mechanical Engineering, Georgia Institute of Technology, USA. This article was published in

1997 under the title 'Incorporating environmental issues in product design and realisation' (Bras, 1997). The article starts with a discussion of the phases in a product's life cycle.

> The concern regarding environmental impact stems from the fact that, whether we want it or not, all our products affect in some way our environment during their life span. In Figure [67] a schematic representation of a system's life cycle is shown. Materials are mined from the earth, air and sea, processed into products and distributed to consumers for usage, as represented by the flow from left to right in the top half of the figure ...
>
> The emergence of product take-back legislation ... has forced manufacturers to think more and more about how to dispose appropriately of their products through recycling and reuse (as represented by the flow from right to left and up in the lower half of [Figure 67]) and to include product retirement issues such as recycling and disassembly as a real requirement in their design process.
>
> (Bras, 1997, p. 7)

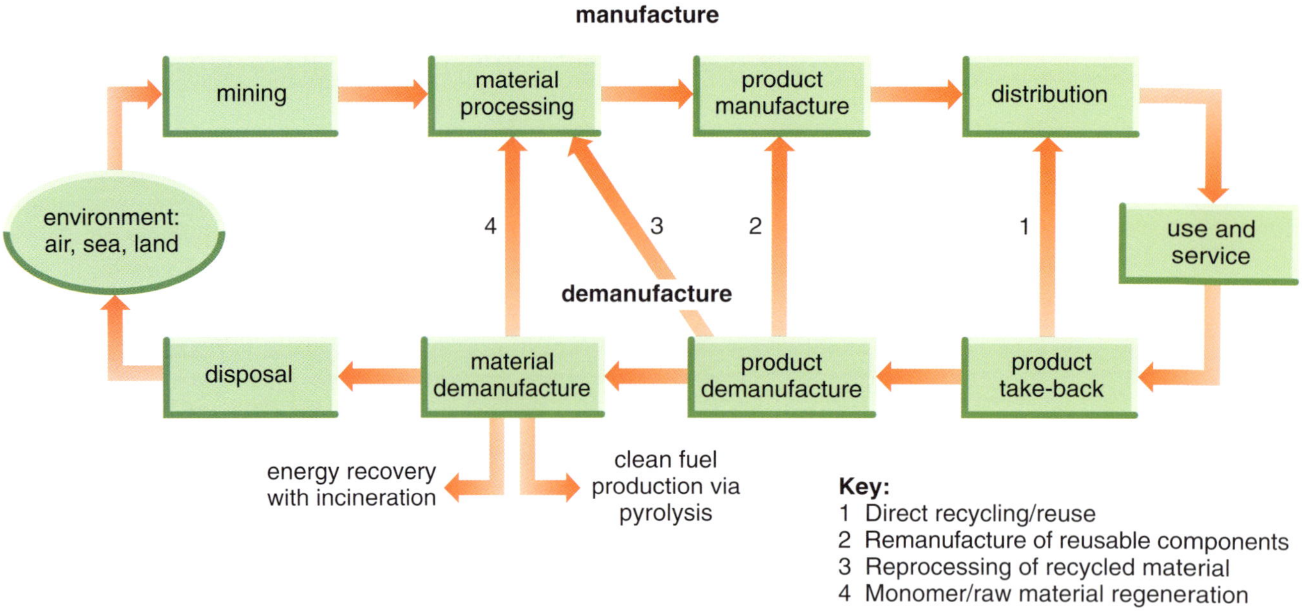

Figure 67 A generic representation of a product's life cycle from cradle to grave and beyond

We have already discussed many of these general issues. Recall that in Block 4 we examined an example of 'take-back' legislation, the End of Life Vehicle (ELV) directive from the European Commission. Other important motivators for companies to become more environmentally responsible are then described:

> - *Legislation*. For example, the US Clean Air Act has limited the use of a number of materials and (European) 'take-back' legislation is driving Design for Recycling efforts.
> - *Customer demand*. Awareness of environmental issues is increasing among customers. Some customers will even pay more for a product if it is 'green'. Also, industrial customers ... do not want (future) environmental liability for a supplier's product.
> - *Eco-labelling programmes*. How 'green' is a product? Having an eco-label becomes a competitive advantage. [For example see Figure 71 of Block 4]

- *ISO 14000*: The ISO 14000 (environmental management standards) certification may become a crucial element in doing business, like ISO 9000 (quality management standards).

(p. 7)

The article then describes the different kinds of approach companies and designers can adopt to reduce environmental impacts. The article presents a useful classification of approaches:

> They can be distinguished by differences in the scope of organisation and temporal concern as shown in Figure [68]. Ideally one would like to move from the lower left corner (arguably the current state of practice) to the upper right corner and achieve sustainable development. In this figure, rather than indicating years, the temporal concern gradations are based on life-spans of products, people and civilisations. The scale as shown is not linear, but is used to indicate important distinctions ... Similarly, the scale of environmental concerns [organisational concern in Figure 68] was chosen to indicate distinctions. These gradations are fairly self-explanatory; although it is worth noting that 'X products' refers to the negative environmental impact of a group of products and that a scope equivalent to 'One manufacturer' would indicate concern about all the activities of a single manufacturing firm, whereas 'X manufacturers' indicates activities among a group of manufacturers. We note that three classes of approaches can be identified, namely:
>
> - those which are applied *within* a single product life cycle and focus on specific life cycle stages;
> - those which focus on a *complete* product life cycle and cover *all* life cycle stages; and
> - those which go *beyond* single product life cycles.
>
> (p. 8)

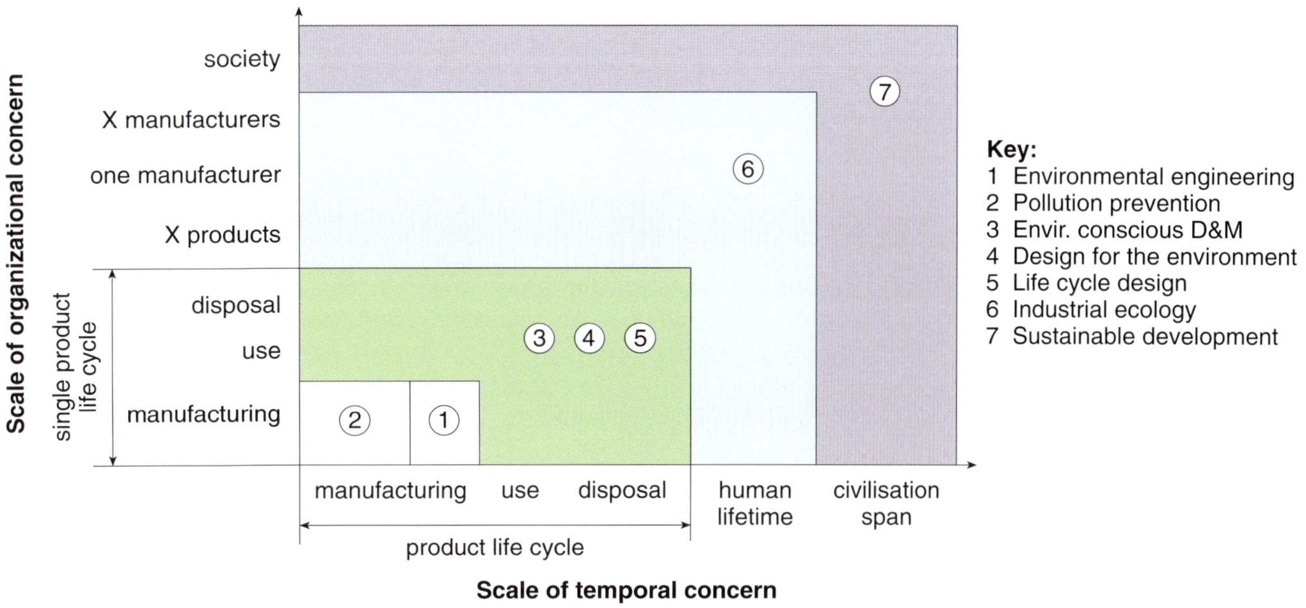

Figure 68 Environmental and temporal impact reduction approaches

In the first class of approaches for reducing environmental impact, ① environmental engineering and ② pollution prevention apply directly to manufacturing and are shown on Figure 68. Other approaches, not named explicitly in Figure 68, also concentrate on specific areas of the life cycle such as 'design for disassembly' or 'design for recycling'.

In the second class, three approaches are identified which relate to whole product life cycles ③ environmentally conscious design and manufacture ④ design for environment and ⑤ life cycle design. The article says that 'all these approaches have similar goals and encourage a holistic product view'.

For the third class, two approaches are mentioned:

⑥ industrial ecology; and
⑦ sustainable development.

We are now well aware of the broad issues in sustainable development, but what is the 'industrial ecology'?

> [I]ndustrial ecology is not limited to a single product life cycle but considers the interactions of several product life cycles (of possibly different lengths) over a larger time scale ... For example, a carpet industry's production waste can be used by car companies to make sound deadening materials.
>
> (p. 9)

This article makes two main points. First, it classifies approaches to sustainable design and secondly, it makes the case that for sustainable design to be more than a generally desirable goal the approaches must be integrated into the process and practices of companies, designers and engineers. As the author comments 'the challenge lies in its seamless implementation in existing and new practices'. Since the article was written this has begun to happen in some larger companies. Philips is an example, publishing annually its independently audited *Sustainability Report*.

The article also examines two types of tools to put these approaches to sustainable design into practice. First, there are guidelines.

> A number of design guidelines, ranging from a focus on specific aspects of a product's life cycle (for example 'reduce the number of materials used to facilitate more efficient recycling') to sustainable development (for example 'reduce the use of non-renewable energy resources') have been developed in recent years ...
>
> As a nice representative example of a relative improvement assessment consider ... LiDS (Lifecycle Design Strategies) wheel in [Figure 69] ... The wheel asks companies to assess their current state with respect to eight life cycle strategies. Note that design guidelines are incorporated as part of the eight strategies in the wheel. Companies can map their current efforts in the spider diagram which highlights where their weak areas are. Once improvements have been made (or planned) a new assessment can be made and overlayed on the previous one, thus showing relative improvements
>
> (p.11)

New concept development: dematerialisation; shared use of the product; integration of functions; functional optimisation of product (components).

Product system level

7 Optimisation of end-of-life system: reuse of product; remanufacturing/refurbishing; recycling of materials; safer incineration.

6 Optimisation of initial lifetime: reliability and durability; easier maintenance and repair; modular product structure; classic design; strong product-user relation.

5 Reduction of impact during use: lower energy consumption; cleaner energy source; fewer consumables needed; cleaner consumables; no waste of energy/consumables.

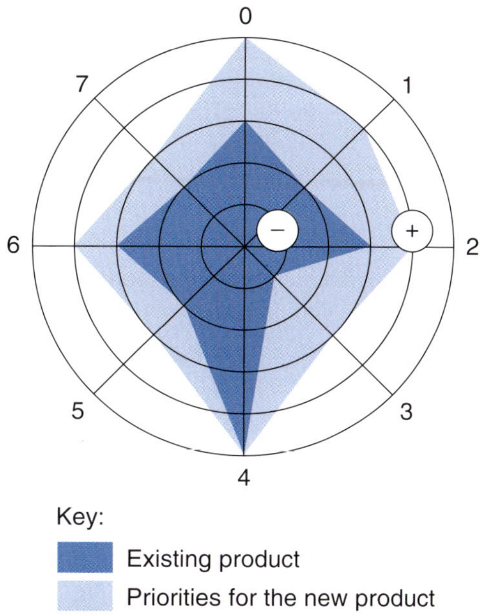

Product component level

1 Selection of low-impact materials: cleaner materials; renewable materials; lower energy content materials; recycled materials; recyclable materials.

2 Reduction of materials usage: reduction in weight; reduction in (transport) volume.

3 Optimisation of production techniques: alternative production techniques; fewer production steps; lower/cleaner energy consumption; less production waste; fewer/cleaner production consumables.

Product structure level

4 Optimisation of distribution system: less/cleaner/reusable packaging; energy-efficient transport mode; energy-efficient logistics.

Figure 69 LiDS (Lifecycle Design Strategies) wheel

(Brezet and van Hemel, 1997, p.81)

Secondly, there are quantitative assessments of the environmental impacts. This is achieved by giving a numerical measure (called a 'quantitative metric') to the impacts.

> Quantitative metrics are needed to assess progress and provide a means for continuous improvement and feedback when integrating environmental issues in design. A variety of metrics have been proposed to allow comparison of design, with most being based on some variant of life cycle assessment (LCA). In many cases a full assessment of the entire environmental costs is not feasible given the work that would be required and the uncertainty of forecasting the changes in the economy and technology over the life of most products.
>
> (p. 11)

The article gives a timely warning about the difficulties in applying quantitative metrics uncritically:

> Even through the number as well as the maturity of different life cycle indices and indicators presented in the literature are rapidly increasing, they still suffer from the fact that a) a lot of data is missing and b) weights and criteria are questionable because non-comparable units of measurement are used, thus leaving a lot of unanswered questions and room for political debate. Additionally, perceptions of environmental

> impact differ between countries ... Nevertheless, many are still very suitable for use in design if one seeks to assess relative product improvements rather that absolute environmental impact.
>
> (p. 11)

The next sub-section presents an example of such a quantitative metric, namely ECO-indicator and its associated software ECO-it.

SAQ 7

What are the advantages of quantitative assessments against design guidelines in sustainable design?

6.4 Quantitative assessment of environmental impacts: ECO-it

The ECO-indicator method, and associated ECO-it ('it' stands for indicator tool) software that you can find on the course DVD, is based on evaluating the damage that a product design might do to the environment at each stage of its life cycle: raw material extraction, manufacture, distribution, use and then at disposal. The first three stages are all concerned with the manufacture of the product and are grouped under the heading 'production'. For the purposes of the ECO-indicator method impacts are divided into just three types: production, use and disposal.

> Don't worry about looking at the ECO-it software yet. There is the chance to do so later in the section.

In order to use the ECO-indicator method to assess a design it is necessary to have accurate information about the materials, methods of manufacture, use and disposal of a product. This is not always easy to obtain. You will need details on the components in the product, including the materials they are made from, how and where they are manufactured and assembled, sources of energy used for manufacture, consumption of energy during use as well as any wastes produced during use. When you have this information the method allows you to calculate impacts.

In this course, you are not expected to assemble this information for even the simplest products. However, you are provided with this information and you will be asked to make assessment of environmental impacts. It is often difficult to obtain accurate information needed to make an ECO-it environmental impact assessment so the results are only estimates.

Before starting I should mention that the impacts are given in terms of indicators. The particular ECO-indicator method uses a scale of impacts measured in 'milliPoints'. This measurement is a common scale for the measurement of all the different types of environmental impacts. For example the pollution impacts, hazardous wastes and energy consumption are all measured in milliPoints.

The indicator values are derived from a complicated combination of different factors including use of natural resources, pollution, energy use, disassembly cost and recycling benefits (negative indicators or positive impacts). All environmental impacts are expressed on a common scale. The *ECO-indicator 99 Manual for designers* (which you can find on the course DVD) explains this scale but basically the main purpose is to compare relative differences between products or components. The scale is chosen in such a way that the value of 1 milliPoint is representative of one millionth of the yearly environmental load of one average European inhabitant.

The relative value of different impacts is important. This allows two things. First, you can make an overall comparison of two designs. Second, you can compare the different aspects of a single design. This can help to indicate which features of a product or phase of its life cycle contribute heavily to environmental impact. This might indicate those areas of the design on which to concentrate redesign effort in order to reduce environmental impacts.

In order to introduce the use of the ECO-indicator method I will outline an example of its application to the design of a coffee maker. This example is presented by the developers of the software in their documentation on the course DVD. The method can be applied either manually by tabulating the results of the analysis or by using the computer program ECO-it to retrieve data on impacts and to make the relevant calculations.

I suggest that at first you follow the manual example. You can then go through the same example using the computer software if you wish. For examples of very simple products there is little advantage in using the computer software.

Example Eco-indicator assessment of a coffee maker

There are five steps in the procedure which is basically the same whether the method is applied manually or with the software. I will concentrate on the manual method for evaluating the coffee maker. The use of the software is explained in the *Block 5 Supplement*.

Step 1: Identify purpose of the assessment

Identify the purpose of the environmental impact assessment. This may be to compare different design proposals or to identify any areas of one design where the impacts can be reduced. It is this latter purpose which guides this assessment of the coffee maker.

Step 2: Define the product life cycle

Describe the main parts of the processes of production, use and disposal. The ECO-it developers use a diagram (see Figure 70 overleaf which refers specifically to the coffee maker) with materials at the top level, which feed into manufacturing processes, transportation, assembly and packaging at lower levels. All materials used throughout the whole life cycle are shown at this top level, including coffee beans and paper (for filters), which are not part of the coffee maker, but are required to use it. Water and electricity are also required to use the coffee maker. Finally, at the bottom level, the items for disposal are identified, i.e. coffee and filters after each use of the coffee maker, packaging and the coffee maker itself at the end of its life.

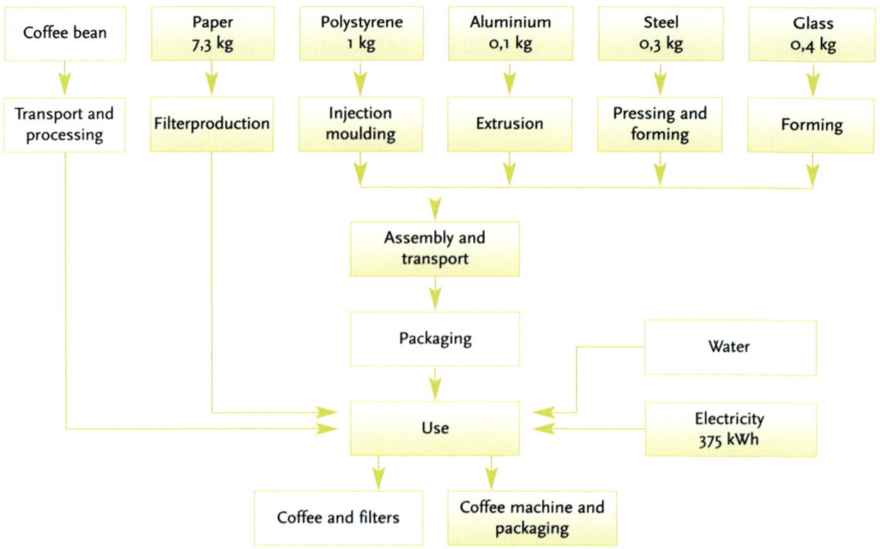

Figure 70 Process tree for a simplified coffee making machine showing material quantities and assumptions

A simplified model of a coffee maker is used in which only the polystyrene housing, the glass jug, the steel hot plate and an aluminium riser pipe are included (the mains cable and the switch are omitted from the example). The items shown as white blocks in Figure 70 are not included in the Eco-indicator calculation in Stage 3. The consumption of coffee and water is omitted because it is difficult for the designer to influence this. The packaging is omitted because this is not under study at this stage.

Step 3: Quantify materials and the processes used in manufacture

Materials, resources and processes are specified and measured. For example the amounts of materials of each type in kilograms are derived from the design description. The energy consumption (kilowatt hours) is based on assumptions: the machine is used twice a day for five years at half capacity (5 cups); the coffee is kept hot for half an hour after it is ready. Assumptions about materials are also made. For example, the glass for the coffee maker is assumed to be packaging glass with a 55% recycled content. The information that is entered into ECO-it takes a well defined form.

(a) Materials used are measured in kilograms. For the coffee maker there is 1 kg of polystyrene, 0.1 kg aluminium, 0.3 kg steel and 0.4 kg of glass.

(b) Manufacturing (processing) of materials is measured either per kilogram or in terms of the shapes produced (areas or volumes). Three main processes are used in the coffee maker.

   (i) The polystyrene casing (1 kg) is injection moulded.

   (ii) The aluminium riser pipe (0.1 kg) is extruded

   (iii) The glass (0.4 kg) is melted and formed into shape. The energy used for this process is estimated at approximately 4 mega joules (which is approximately 1.1 kilowatt hours).

   (iv) Processing of the steel hot plate has low environmental impact for this small component and is ignored in the calculation.

(c) Energy is measured in kilowatt hours (or mega joules for heat energy). Electricity use is estimated at 375 kWh in total for two uses per day over five years.

(d) Transport is measured by tonnes (1000 kg) transported one kilometre. No allowances for transport are included in the analysis of the coffee maker.

(e) Recycling and treatment are measured per kilogram of material submitted for processing. For the coffee maker assumptions are made about how the user disposes of it. It is unlikely that the machine will be dismantled and disposed of separately in different collection systems. We assume that the machine will be put in the dustbin and thus processed as municipal waste. Note that in the table (Figure 71) for disposal the aluminium and steel is put together as 0.4 kg of ferrous material (although of course this is not strictly accurate as only the steel is ferrous). However, the glass jug, provided it is designed such that it will fit through the opening of a glass recycling container, can be regarded as household waste. This assumes that a certain proportion of households dispose of glass in a glass recycling container.

Step 4: Calculate impacts

Environmental impacts are then calculated by reference to data tables containing indicator values for the impacts of different amounts of materials and resources as well as different processes, modes of use and disposal. These data tables can be found at the end of the *ECO-indicator 99 Manual for designers* (on the course DVD). I do not show these full data tables here in the text. However, the results of consulting the tables for the coffee maker are given in Figure 71 overleaf.

The material indicator value for 1 kg of clear glass is 58 milliPoints, giving 23 milliPoints for 0.4 kg. The 4 MJ of heat has an indicator value of 21 milliPoints. Disposal of 0.4 kg of glass has indicator value of -2.8 milliPoints. The negative value occurs because recycling the glass as household waste reduces its environmental impact. The material indicator value of 1 kg of polystyrene is 360 milliPoints. The indicator for electricity supply in the Netherlands is used in the assessment of the coffee maker example because the Eco-indicator method is developed in the Netherlands.

Step 5: Interpret results of assessment

The last stage is interpreting the results of the calculations. ECO-indicator assessment of the coffee maker indicates the overwhelming environmental impacts are from energy consumed heating the coffee during use and then from one-use paper filters. The ECO-indicator results can be used to identify areas where the design can be improved. Designers for a new coffee maker might consider a thermos jug instead of a hot plate for keeping the coffee hot. In addition, a permanent filter could replace single-use paper filters.

I suggest that you now open the ECO-it software and turn to the section titled ECO-it in the *Block 5 Supplement*. If it's not convenient to access your computer at this point return to it as soon as you can.

The ECO-it activity in the *Block 5 Supplement* suggests that looking up indicator values and making calculations for an assessment can be labourious. However, the great advantage of using the ECO-it software is that it helps you look up the indicator values and it does all the calculations automatically!

Product or component	*Project*
coffee machine	**example**
Date	*Author*
14-4-2000	**PRé**
Notes and conclusions	
Analysis of a coffee machine, assumption: 5 years' use, 2 x per day, half capacity, keep hot for 30 minutes	

Production (Materials, treatments, transport and extra energy)			
material or process	**amount**	**indicator**	**result**
polystyrene	1 kg	360	360
injection moulding PS	1 kg	21	21
aluminium	0,1 kg	780	78
extrusion Al	0,1 kg	72	7
steel	0,3 kg	86	26
glass	0,4 kg	58	23
gas-fired heat (forming)	4 MJ	5,3	21
Total [mPt]			536

Use (Transport, energy and possible auxiliary materials)			
process	**amount**	**indicator**	**result**
electricity	375 kWh	37	13.875
low-voltage			
paper	7,3 kg	96	701
Total [mPt]			14.576

Disposal (Disposal processes for each material type)			
material and type of processing	**amount**	**indicator**	**result**
municipal waste, PS	1 kg	2	2
municipal waste, ferrous	0,4 kg	-5,9	-2,4
household waste, glass	0,4 kg	-6,9	-2,8
municipal waste, paper	7,3 kg	0,71	5,2
Total [mPt]			2
Total [mPt] (all phases)			15.114

Figure 71 Data tables for the coffee maker

Assessing environmental impacts throughout a product's life cycle is an important aspect of sustainable design. As you have seen the assessment relies on good data about the materials, processes, modes of use and disposal of a product. In many cases such data can be uncertain and approximate and is the subject of ongoing research.

Key points of Section 6

A design is sustainable if it meets a wide spectrum of needs – social, environmental and economic. Today, sustainability is a critical function for an increasing number of products. Sustainable development does not compromise the needs of future generations.

Lessening environmental impacts cannot automatically be thought of as sustainable design, it is only a contribution towards it.

The life cycle of a product has five main phases: material extraction, production (manufacture), distribution, use and end of life disposal, recycling or reuse.

Designing for sustainability is difficult partly because a product may have many social, environmental and economic consequences or impacts during its life cycle and partly because these consequences are difficult to predict accurately whilst the product is being designed. A recent report by Philips identifies their responsibilities and responses under these three headings.

Approaches to sustainable design are classified as

1 within a single product life cycle and focus on specific life cycle stages;

2 cover all life cycle stages of individual products; and

3 go beyond single product life cycles.

Just as with other areas of design, modelling can be used. A variety of quantitative assessment tools can be used to model the impacts of a product, e.g the ECO-indicator method. Such tools are increasingly computer-based such as ECO-it. These tools allow

(a) comparison of designs;

(b) progress on sustainability to be assessed; and

(c) evaluation of trade-offs between various features of a design.

Factors which encourage companies to be more aware of sustainability issues, environmental responsibilities legislation, consumer demand, eco-labelling and standards.

(Learning outcomes: 1.5, 2.4, 3.3, 4.1)

Answers to self-assessment questions

SAQ 1

What are the main characteristics of Thonet chairs which we can think of as belonging to modern mass production?

Firstly, the products are standardised. Secondly, the work processes were also standardised in the sense that they were repetitive and increasingly mechanised. Thirdly, the use of standard fixings and precision formers meant that component parts were interchangeable. In packing chairs for dispatch, each piece, say a chair back, was not a unique individual piece to be fitted to other unique individual components, say a chair seat. Any one type of piece could stand in for other pieces of the same type.

SAQ 2

Traditional 'dolly' pegs, shown in Figure 26(b), have been around for centuries. Speculate on the factors that made this type of clothes peg commercially viable when modern pegs must exploit mass production to be commercially viable.

Dolly pegs are a product of a rural cottage industry that provided a very low but steady income for its workers. It was a relatively low-skill technique whereby each peg was carved to an approximate form and bound with twine, wire, metal or, more recently, short pieces of tubing, to resist splitting. Overheads would have been low because raw materials (typically wood from willows) were available locally and only a few basic tools were required. Much time would have gone into the production and local distribution for a very meagre income.

SAQ 3

We've come across plywood several times in the course. It's discussed at length in Block 1. Where would you place it in the hierarchical materials classification presented above?

Plywood consists of thin veneers of natural wood held together with various forms of adhesive. It is therefore a composite material. Sometimes a hardwearing plastic laminate is applied to one surface making it ideal for use as a worksurface. Plywood is best referred to as a laminar composite.

SAQ 4

What do you think might be the most appropriate type(s) of production system for making the following products?

- domestic refrigerator

flow-line

- racing bicycle

bespoke, batch or flow-line depending on volume required

- agricultural fertiliser

continuous

- hi-fi amplifier

batch or flow-line depending on volume required

- popular model of a family car

flow-line

- dining chairs sold by a major out-of-town warehouse chain

batch (or possibly flow-line)

- a unique sculpture commissioned from an artist

one-off

SAQ 5

Can you recall from Section 3 some examples of processes which alter materials properties?

Steaming can make some timbers pliable.

Heating can soften some polymers.

Processes involving heating can alter the properties of some metals i.e. they can be made stronger, harder or tougher.

Work hardening of some metals occurs as a result of processes such as cold rolling or bending.

SAQ 6

Outline briefly the three main areas of responsibility that the Philips *Sustainability Report* describes.

Economic responsibility is ensuring that the company as a whole is sustainable financially. *Environmental responsibility* is about reducing environmental impacts of the products in terms of the processes of production, product use and disposal. *Social responsibility* concerns the health, safety and education of the workforce as well as effects on people outside the organisation.

SAQ 7

What are the advantages of quantitative assessments against design guidelines in sustainable design?

They allow:

(a) comparison of designs;

(b) progress on sustainability to be assessed; and

(c) evaluation of trade-offs between various features of a design.

Appendix A
Widely used engineering materials, with short names

Metals

Aluminium alloys: Al-alloys

Cast irons

Copper alloys: Cu-alloys

Lead alloys: Pb-alloys

Magnesium alloys: Mg-alloys

Nickel alloys: Ni-alloys

Steel – carbon: Steel

Steel – low alloy: LA steel

Steel – stainless: SS

Titanium alloys: Ti-alloys

Tungsten alloys: W-alloys

Zinc alloys: Zn-alloys

Ceramics and Glasses

Alumina: Al_2O_3

Aluminium Nitride: AlN

Boron Carbide: BC

Borosilicate glass

Brick

Concrete

Glass ceramic

Silica glass: SiO_2

Silicon

Silicon Carbide: SiC

Silicon Nitride: Si_3N_4

Soda-Lime glass

Stone

Tungsten Carbides: WC

Natural materials

Bamboo

Cork

Leather

Wood

Polymers and elastomers

Acrylonitrile butadiene styrene: ABS

Butyl Rubber: BR

Cellulose polymers: CA

Epoxies

Ethyl vinyl acetate: EVA

Ionomer: I

Isoprene: IR

Natural Rubber: NR

Phenolics

Polyamides: Nylons, PA

Polycarbonate: PC

Polychloroprene: Neoprene, CR

Polyester

Polyetheretherketone: PEEK

Polyethylene terephalate: PET or PETE

Polyethylene: PE

Polymethyl methacrylate: Acrylic, PMMA

Polyoxymethylene: Acetal, POM

Polypropylene: PP

Polystyrene: PS

Polytetrafluroethylene: PTFE

Polyurethane: PUR

Polyvinylchloride: PVC

Silicone elastomers: SIL

Composites

Aluminium/Silicon Carbide Composite: Al/SiC

Carbon fibre reinforced polymers: CFRP

Glass fibre reinforced polymers: GFRP

Foams

Flexible Polymer Foam

Rigid Polymer Foam

References and further reading

References

Ashby, M.F. and Johnson K. (2002) *Materials and Design – the Art and Science of Materials Selection in Product Design*, Oxford, Butterworth Heinemann.

Bras, B. (1997) 'Incorporating environmental issues in product design and realisation', *Industry and Environment*, vol. 20, no. 1–2, pp. 7–13.

Brezet, H. and Van Hemel, C. (eds.) (1997) *Ecodesign: A promising approach to sustainable production and consumption*, Paris, United Nations Publications.

Brundtland, G. (ed.) (1987) *Our Common Future: The World Commission on Environment and Development*, Oxford, Oxford University Press.

Candilis G., *et al.*, (1980) Bugholzmöbel, Stuttgart, Katl Kramer Verlag,

Dagger, B. K. J. (1985) Design for Economic Manufacture, Engineering Industry Training Board.

Platts, J. (1990) 'Design and manufacture', unpublished paper presented at Design Seminar, Faculty of Technology, The Open University, June.

Philips (2003) *Sustainability Report 2002*, Royal Philips Electronics, Eindhoven. (A copy of this is available on the course DVD)

Wilk, C. (1980) *Thonet: 150 Years of Furniture*, Barrons.

Womack, J.P., Jones, D.T. and Roos, D. (1990) *The Machine that Changed the World*, Rawson Associates/Macmillan.

Further reading for Section 2

Prepared by Mike Ashby

Ashby, M.F. and Johnson K. (2002) *Materials and Design – the Art and Science of Materials Selection in Product Design*, Oxford, Butterworth Heinemann.
(A book that develops further the ideas outlined in Section 2.)

Clark, P. and Freeman, J. (2000) *Design, a Crash Course*, New York, NY, USA, The Ivy Press Ltd, Watson-Guptil Publications, BPI Communications Inc.
(An entertainingly-written scoot through the history of product design from 5000 BC to the present day.)

Dormer, P. (1993) *Design since 1945*, Thames and Hudson, London.
(A well-illustrated and inexpensive paperback documenting the influence of industrial design in furniture, appliances and textiles – a history of contemporary design that complements the wider-ranging history of Haufe (1998), see below.)

Forty, A. (1986) *Objects of Desire – Design in Society since 1750*, London Thames and Hudson.
(A refreshing survey of the design history of printed fabrics, domestic products, office equipment and transport system. The book is mercifully free of eulogies about designers, and focuses on what industrial design does, rather than who did it. The black and white illustrations are disappointing, mostly drawn from the late 19th or early 20th centuries, with few examples of contemporary design.)

Haufe, T. (1998) *Design, a Concise History*, London, Laurence King Publishing (originally published in German).
(An inexpensive soft-cover publication. Probably the best introduction to industrial design for students (and anyone else). Concise, comprehensive, clear and with intelligible layout and good, if small, colour illustrations.)

Jordan, P.S. (2000) *Designing Pleasurable Products*, Taylor and Francis, London, UK.
(Jordan, Manager of Aesthetic Research and Philips Design, argues that products today must function properly, must be usable, and must also give pleasure. Much of the book is a description of market-research methods for eliciting reactions to products from users.)

Julier, G. (1993) *Encyclopedia of 20th Century Design and Designers*, London, Thames and Hudson.
(A brief summary of design history with good pictures and discussions of the evolution of product designs.)

Manzini, E. (1989) *The Material of Invention*, London, The Design Council.
(Intriguing descriptions of the role of material in design and in inventions. The translation from Italian to English provides interesting – and often inspiring – commentary and vocabulary that is rarely used in traditional writings about materials.)

McDermott, C. (1999) *The Product Book*, UK, D & AD in association with Rotovison.
(50 essays by respected designers who describe their definition of design, the role of their respective companies and their approach to product design.)

Norman, D.A. (1988) *The Design of Everyday* Things, New York, Doubleday.
(A book that provides insight into the design of products with particular emphasis on ergonomics and ease of use.)

Acknowledgements

The Course Team would like to acknowledge the contribution made to this block by material originally developed for T204 *Design Principles and Practice* Block 5 *Product Development and Manufacture* by David Walker, Robin Roy and Barry Dagger.

Grateful acknowledgement is made to the following sources for permission to reproduce material within this book:

Figures 1, 3, 5, 6 and 8: Gebrüder Thonet GmbH; Figure 2: Esprit Nouveau Pavilion 1925, Le Corbusier, © DACS 1993, photo Lucien Hervé; Figure 7: Vitra Design Museum – Archive Alexander von Vegesack; Figure 11: www.vintagepens.com; Figure 19: Bang & Olufsen; Figure 20: Top right: http://www.incx.nec.co.jp/robot; top centre and lower left: Victoria and Albert Museum; top left: © Scott L. Robertson/slrobertson.com; lower right: Zanotta Spa; lower centre: © Bauhaus Dessau Foundation; Figure 23: Fuji Photo Film (UK) Ltd; Figure 24: The Concrete Centre; Figure 25: ALPA Capaul & Weber; Figure 35: The Motoring Picture Library, Beaulieu

Every effort has been made to trace all copyright owners, but if any has been inadvertently overlooked, the publishers will be pleased to make the necessary arrangements at the first opportunity.

Course Team

Academic staff

Ken Baynes, External Assessor
Catherine Cooke, Author
Nigel Cross, Author
Chris Earl, Author
Steve Garner, Author and Course Chair
Georgy Holden, Author
Robin Roy, Author

Consultants

Mike Ashby, Contributing Author, Block 5
Mark Evans, Contributing Author, Workbook 1
Tony Hodgson, Contributing Author, Block 5

Associate lecturers

Jenny Burke
Nick Jeffrey

Course managers

Andy Harding, Course Manager
Amber Thomas, Course Manager

Production staff

Tammy Alexander, Graphic Designer
Margaret Barnes, Course Secretary
Deirdre Bethune, Course Secretary
Philippa Broadbent, Print Buyer
Jane Bromley, Interactive Media Designer
Michael Brown, Video Editor
Daphne Cross, Assistant Print Buyer
Tony Duggan, Learning Projects Manager
Bernie D'Souza, Course Secretary
Vicky Eves, Graphic Artist
Barbara Fraser, Picture Researcher
Phil Gauron, Video Producer
Richard Hearne, Photographer
Lynda Jones, Rights Assistant
Jane Moore, Editor
Jonathon Owen, Graphic Artist
Val Price, Rights Executive
Alex Reid, Narrator (video)
Ekkehard Thumm, Media Project Manager
Robert Wood, Editor